STAINLESS STEELS AND ALLOYS

Edited by **Zoia Duriagina**

Stainless Steels and Alloys

http://dx.doi.org/10.5772/intechopen.76956
Edited by Zoia Duriagina

Contributors

Tatsuhiko Aizawa, Shiyun Dong, Xiang-Yi Feng, Jin Xiang Fang, Shixing Yan, Kamran Saeidi, Farid Akhtar, Adriana Da Cunha Rocha, Andrea Pedroza Da Rocha Santos, Gabriela Ribeiro Pereira, Ramesh Rudrapati, Jose Daniel Biasoli De Mello, Willian Silva Labiapari, Miguel Angel Narvaez Ardila, Henara Lilian Costa, Shamanth V, Hemanth K, Ravishankar K S, Yong A Zhang, Rui Xuan Li

Notice

First published in London, United Kingdom, 2019 by IntechOpen
IntechOpen is the global imprint of INTECHOPEN LIMITED, registered in England and Wales, registration number: 11086078, The Shard, 25th floor, 32 London Bridge Street
London, SE19SG – United Kingdom
Printed in Croatia

British Library Cataloguing-in-Publication Data
A catalogue record for this book is available from the British Library

Additional hard copies can be obtained from orders@intechopen.com

Stainless Steels and Alloys, Edited by Zoia Duriagina
p. cm.
Print ISBN 978-1-78985-369-8
Online ISBN 978-1-78985-370-4

For EU product safety concerns:
IN TECH d.o.o., Prolaz Marije Krucifikse Kozulić 3, 51000 Rijeka, Croatia,
info@intechopen.com or visit our website at intechopen.com.

Meet the editor

Prof. Zoia Duriagina, DrSc, PhD, DrHabEng, is a full professor of materials science and materials engineering. She is Head of the Department of Applied Materials Science and Materials Engineering of Lviv Polytechnic National University (Ukraine) and Professor of the Department of Physical Chemistry and Physico-Chemical Fundamentals of Environmental Engineering of the John Paul II Catholic University of Lublin (Poland). She was awarded the golden medal of the World Academy of Materials Science and Manufacturing Engineering. Her main area of study is the relationship between the nano- and microstructure of materials and their properties. She is the author of about 295 publications, including 10 books and 18 patents. As a visiting professor, he lectures at universities in Poland, Belgium, and Germany.

Contents

Preface

It is with great pleasure that I have been asked to edit the book *Stainless Steels and Alloys*. This topic is very familiar to me because it has been the main area of my scientific interest for a long time—from my student years when I was working on my graduation work dedicated to the study of the properties of stainless steels for surgical instruments. It was then that I realized that materials science is the magic that allows us to the change the chemical composition and microstructure of material to regulate its corrosion-mechanical, technological, and functional properties. Later, both of my theses were related to the study of the corrosion resistance of stainless steels and the development of methods for applying protective coatings. I have already published many studies as well as books, monographs, and numerous papers on this topic.

There is a widely held view that stainless steel was discovered in 1913 by Sheffield metallurgist Harry Brearley. He was experimenting with different types of steel for weapons and noticed that a 13% chromium steel had not corroded after several months. Although there is much mystery and speculation behind the discovery of this wonderful material, there is no question that without the help of many scientists and metallurgists we would not have such a rich and versatile metal at our fingertips.

The most common forms of corrosion in stainless steel are:

Pitting corrosion— The passive layer on stainless steel can be attacked by certain chemical species. The chloride ion Cl^- is the most common of these and is found in everyday materials such as salt and bleach.

Crevice corrosion—Stainless steel requires a supply of oxygen to make sure that the passive layer can form on the surface. In very tight crevices, it is not always possible for oxygen to gain access to the stainless steel surface thereby causing it to be vulnerable to attack.

General corrosion—Normally, stainless steel does not corrode uniformly as do ordinary carbon and alloy steels. However, with some chemicals, notably acids, the passive layer may be attacked uniformly depending on the concentration and temperature, and metal loss is distributed over the entire surface of the steel.

Stress corrosion cracking—This is a relatively rare form of corrosion that requires a very specific combination of tensile stress, temperature, and corrosive species, often the chloride ion, for it to occur.

Intergranular corrosion—This is now quite a rare form of corrosion. If the carbon level in the steel is too high, chromium can combine with carbon to form chromium carbide. This process is also called sensitization and typically occurs during welding. The chromium available to form the passive layer is effectively reduced and corrosion can occur.

Galvanic corrosion—If two dissimilar metals are in contact with each other and with an electrolyte, e.g., water or other solution, it is possible for a galvanic cell to be set up.

Five major classes of stainless steels are widely used: ferritic, austenitic, martensitic, duplex, and precipitation hardening. Austenitic stainless steels are extensively used for service down to as low as the temperature of liquid helium (–269ºC). This is largely due to the lack of a clearly defined transition from ductile to brittle fracture in impact toughness testing. Steels with ferritic or martensitic structures show a sudden change from ductile (safe) to brittle (unsafe) fracture over a small temperature difference. Even the best of these steels shows this behavior at temperatures higher than –100ºC and in many cases only just below zero. Various types of stainless steel are used across the whole temperature range from ambient to 1100ºC. In the European standards, a distinction is made between stainless steels and heat-resisting steels. However, this distinction is often blurred and it is useful to consider them as one range of steels.

I would like to thank all my coauthors and cooperators from IntechOpen for their efforts in preparing this book. I hope that the new experimental data given in the book will be the foundation for the development of new materials and technologies in materials science. This book will be useful to scientists, engineers, masters, graduate students, and students. I hope readers will enjoy this book and that it will serve to create new materials with unique properties.

Prof. Zoia Duriagina, DrSc, PhD, DrHabEng
Head of Department of Applied Materials
Science and Materials Engineering of
Lviv Polytechnic National University, Ukraine
Prof. of the John Paul II Catholic University of Lublin, Poland

Chapter 1

Phase Transformations in Duplex Stainless Steel: An Assessment by In Situ X-Ray Diffraction

Adriana da Cunha Rocha,
Andrea Pedroza da Rocha Santos and
Gabriela Ribeiro Pereira

Additional information is available at the end of the chapter

http://dx.doi.org/10.5772/intechopen.81128

Abstract

Duplex stainless steels (commonly known as DSS) are a class of stainless steels with a microstructure formed by two main phases: ferrite and austenite. They are used in a wide range of applications, such as chemical processing, in maritime environments and in the oil and gas industries. In most cases, DSS are chosen based on their strength and corrosion resistance for various environments. When exposed to temperatures above 600°C though, the balance of alloying elements can be modified due to precipitation of various secondary phases, such as sigma (σ) and chi (χ). The sigma phase is typically enriched with Cr and Mo, so its formation can lead to a drastic deterioration in toughness, corrosion resistance, and weldability of duplex stainless steels. To prevent damages to these steels due to the formation of sigma phase, the understanding of such transformation becomes mandatory, not only during the development of these steels but also during their processing. In this research, samples from a lean duplex steel UNS S32304 are subjected to a temperature of 800°C and analyzed in situ by X-ray diffraction. Thus, the kinetics of phase transformations occurring in duplex stainless steels are observed in real time.

Keywords: duplex stainless steels, sigma phase, X-ray diffraction, in situ measurements, lean duplex steel

IntechOpen

1. An introduction to duplex stainless steels

1.1. Historical background

The first-generation duplex stainless steels, produced in the 1930s, had a satisfactory performance and have been used for a variety of process industry applications including vessels, heat exchangers, and pumps but had limitations in the as-welded condition because of its excessive ferrite content. The introduction of new refining processes [such as the argon oxygen decarburization (AOD)] opened the possibility of a broad spectrum of new stainless steels. Also, the addition of nitrogen as an alloying element made it possible to increase toughness and corrosion of the heated affected zone resistance approaching that of the base metal in the as-welded condition.

The second-generation duplex stainless steels were defined by their nitrogen alloying. This new commercial development, which began in the late 1970s, coincided with the development of offshore gas and oil fields in the North Sea and the demand for stainless steels with excellent chloride corrosion resistance, good fabricability, and high strength. Duplex 2205 stainless steel became the workhorse of the second-generation duplex grades and was used extensively for gas gathering line pipe and process applications on offshore platforms. The high strength of these steels allowed for reduced wall thickness and reduced weight on the platforms and provided considerable incentive for their use [1].

Recent developments in stainless steels are driven, among other factors, by the low cost of the final material, which led to the evolution of the duplex stainless steel subgroup known as "lean duplex." The main objective is the development of DSS with lower content of alloying elements than standard duplex stainless steels [2]. They present as typical chemical composition 20–24% of chromium, 1–5% of nickel, 0.1–0.3% of molybdenum, and 0.10–0.22% of nitrogen. Besides that, its mechanical properties are twice as great as that of austenitic steels, and it has better corrosion resistance and higher yield strength allowing a reduction in the thickness of many final products [3].

1.2. Alloying elements and secondary phases

The constant evolution of stainless steels resulted in complex compositions containing significant alloying elements, such as chromium, nitrogen, nickel, and molybdenum. To achieve a stable duplex structure that responds well to processing and fabrication, care must be taken to obtain the correct level of each of these elements, since adding them to the steel will affect its mechanical and corrosion properties [4, 5].

The benefits of adding alloying elements are attached to unavoidable disadvantages, the main one being a microstructural instability of the material. During processing or use, duplex stainless steels are subject to various phase transformations caused by temperature and time variations, leading to precipitation of various secondary phases. The most common are precipitated austenite, nitrides, carbides (mostly $M_{23}C_6$ type), and intermetallic elements phases, such as sigma (σ) and chi (χ). The formation of such phases is undesirable as they may cause a significant decrease in corrosion resistance and reduction of the mechanical properties of the material. Therefore, careful processing is necessary to avoid or at least to minimize such transformations [6, 7].

The schematic time-temperature-transformation (TTT) diagram from **Figure 1** shows the typical ranges of temperature and time for precipitation of principal phases in duplex stainless steels. It is observed that the precipitation of the principal intermetallic phases occurs in temperatures above 600°C. Care should be specially taken for the formation of sigma phase, because certain amounts of this phase are extremely detrimental for the mechanical and corrosion properties of theses steels. Therefore, previous knowledge of this amount is extremely important in order to assess the steel.

1.3. The sigma phase

Sigma phase is a nonmagnetic intermetallic phase, rich in iron, chromium, and sometimes molybdenum that presents a complex tetragonal crystalline structure [8]. Its presence affects negatively the mechanical properties, corrosion resistance, and weldability of duplex stainless steels. Precipitation of sigma phase occurs in the duplex stainless steels when these are subjected to high temperatures, either by casting, welding, forging, and aging, as reported by many authors [9–11].

Villanueva et al. [12] successfully showed that sigma phase precipitates preferentially on the austenite/ferrite grain boundaries (**Figure 2**), consuming the ferrite-forming elements (iron and chromium) and growing toward this phase. Similar reports were made by Magnabosco [13], where it is observed that sigma phase precipitation is between 600 and 1000°C, the fastest precipitation rate is occurring around 800–850°C. *Sigma* phase can precipitate by three distinct mechanisms: nucleation and growth from the original ferrite, eutectoid transformation of ferrite into secondary austenite and sigma, and growth from austenite after the total consumption of original ferrite. In any case, the 800°C temperature range is the one to be observed [13, 14].

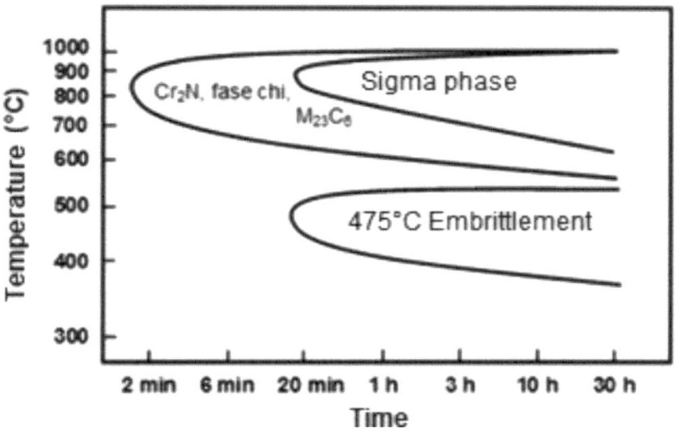

Figure 1. Kinetics of precipitation of different phases in duplex stainless steel [7].

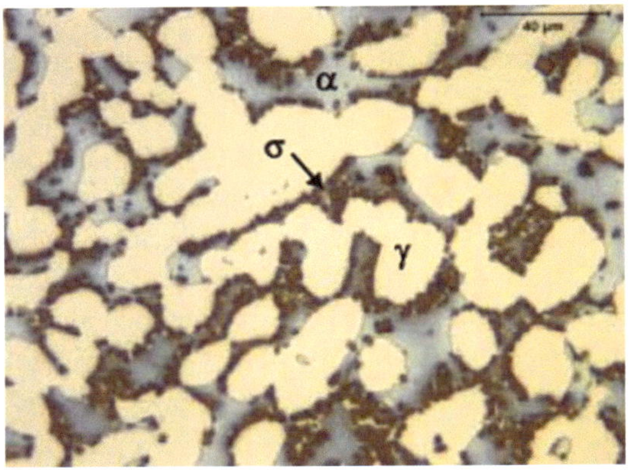

Figure 2. Optical micrograph showing the microstructure of the sample after the heat-treating cycle. Ferrite (α), austenite (γ), and sigma (σ) phase are indicated [10].

2. X-ray diffraction

X-ray diffraction (XRD) is an extremely important analytical technique in the field of microstructural characterization of materials, allowing to obtain information of the order of 10^{-8} cm (1 Å) [15]. Its applications include identification and characterization of crystalline materials, determination of crystal structures, and quantification of phases, both using Rietveld method, textural measurements, residual stress, among others.

Data acquisition to study kinetics at non-ambient conditions is possible with a combination of the XRD equipment with a temperature chamber. X-ray diffraction at non-ambient conditions can be used for a variety of applications, including the study of dynamic processes that need to be investigated in situ, such as phase transformations, oxidation reactions, and crystallite growth.

3. The Rietveld method

The Rietveld method is a mathematical method that performs an adjustment of data on crystalline structures of materials analyzed by X-ray diffraction. In this method, least-square refinements are carried out until the best fit is obtained between the entire observed powder diffraction pattern and the full calculated pattern, as represented in **Figure 3**. The method itself performs the adjustment of data taking into account the total scan and not only one phase of the material [16, 17].

The quantity minimized in the least-square refinement is the residual s_y and is calculated according to Eq. (1):

$$S_y = \sum_i W_i (y_i - y_{ci})^2 \tag{1}$$

where $w_i = 1/y\_i$, y_i = observed intensity at the i-th step, and y_{ci} = calculated intensity at the i-th step.

The method's basic premise is that no efforts should be made in advance to allocate observed intensity to particular Bragg reflections or to resolve overlapped reflections. Therefore, it is necessary that a good starting model is used. In this case, different Bragg reflections contributing to a specific intensity y_i are taken into account at every specific i point in the whole pattern. The calculation of intensities can be performed considering the structure factor F_k, for the K_{th} Bragg reflection. This is done by summing the calculated contributions from neighboring Bragg reflections (within a specific range) plus the background, as described in Eq. (2):

$$y_{ci} = s \sum_k L_k F_k^2 \Phi(2\theta_i - 2\theta_k) P_k A + y_{bi} \tag{2}$$

where y_{ci} = calculated intensity of radiation in the i-th step, s = scale factor, k = represents the Miller indices hkl for a Bragg reflection, L_k = contains the Lorentz polarization and multiplication factors, F_k = structure factor modulus for the K_{th} Bragg reflection, Φ = reflection profile function, P_k = preferred orientation function, A = material absorption factor, and y_{bi} = background intensity at the i-th step.

In the quantitative phase analysis using the Rietveld method, the relative weight fraction W of each phase p in a mixture of n phases is calculated according to Eq. (3):

$$W_p = \frac{S_p (ZMV)_p}{\sum_{j=1}^{N} S_j (ZMV)_j} \tag{3}$$

where W_p = weight fraction of phase p, S = Rietveld scale factor, Z = number of formula units per unit cell, M = mass of the formula unit (in atomic mass units), and V = unit cell volume.

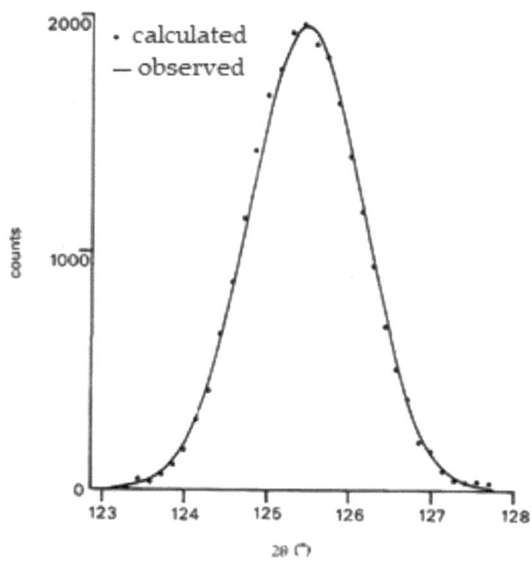

Figure 3. Adjustment by the Rietveld method [17].

4. Experimental procedures

4.1. Chemical composition of steel alloys

Samples with dimensions 20 mm × 20 mm × 1.5 mm were taken from a lean duplex steel plate UNS S32304, with chemical composition presented in **Table 1**.

4.2. In situ high-temperature X-ray diffraction

In situ high-temperature X-ray diffraction data were collected using a *Bruker D8 Discover* diffractometer with a domed hot stage *Anton Paar-DHS 900* (temperature range of 25–900°C), depicted in **Figure 4**. The source of radiation used was cobalt, with a wavelength of 1789 Å.

Diffraction patterns were initially collected at room temperature (30°C). Samples were then heated at 800°C, with an average rate of 1°C/s, and remain at this temperature for 30, 60, or 90 min. At 800°C, the diffraction patterns were collected at every 10 min. At the end of each time interval, cooling was performed at an average rate of 1°C/s till it reaches room temperature.

The main phases were identified by indexing the peaks in the EVA program using the diffraction patterns database provided by PDF Maint. Quantitative phase analysis of the samples was performed using the Rietveld method [16, 17] aided by Topas software.

4.3. OM and SEM-EDS characterization

Metallographic analyses were performed under a *Zeiss* optical microscope (OM), using the AxioVision Rel. 4.7 program, using the light field reflection technique. To obtain a better

C	Co	Cu	Cr	Fe	Mn	Mo	N	Ni	Si
0.036	0.128	0.418	22.670	70.000	1.290	0.117	0.110	4.640	0.385

Table 1. Chemical composition of the lean duplex steel (%wt).

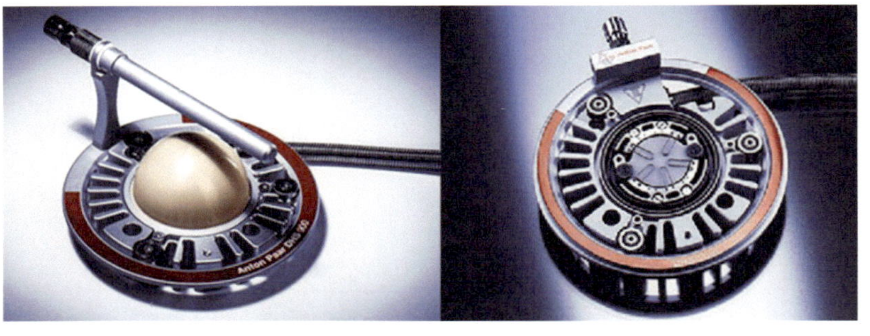

Figure 4. Oven coupled to the X-ray diffractometer.

differentiation between ferrite/austenite and the secondary phases formed, an electrolytic attack with 40% NaOH solution was carried out, with 0.2 V for 300 s. Scanning electron microscopy (SEM) with backscattered electrons and EDS analysis was also performed.

5. Results

5.1. Microstructure

Optical microscopy analysis revealed the existence of three distinct regions (white, gray, and brown) in all heated samples, as observed in **Figures 5–7**. The three regions found in the micrographs can be related to presence of the ferrite (δ), austenite (γ), and sigma phase (σ). These observations are in agreement with the results obtained by Jackson et al. [18] for the same selective etching performed in their work, to distinguish these specific phases.

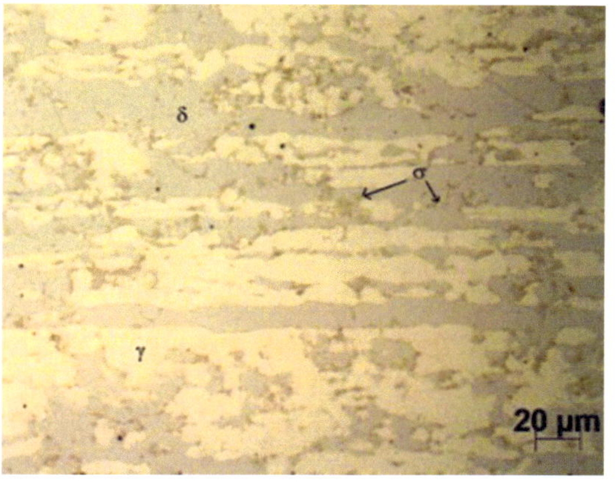

Figure 5. Typical sample after a 30min interval at 800°C.

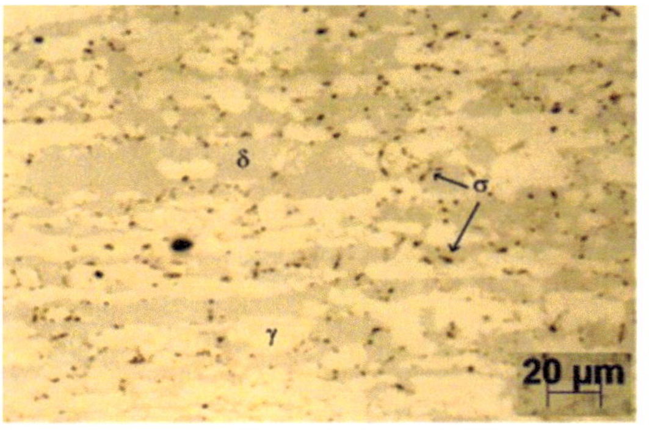

Figure 6. Typical sample after a 60 min interval at 800°C.

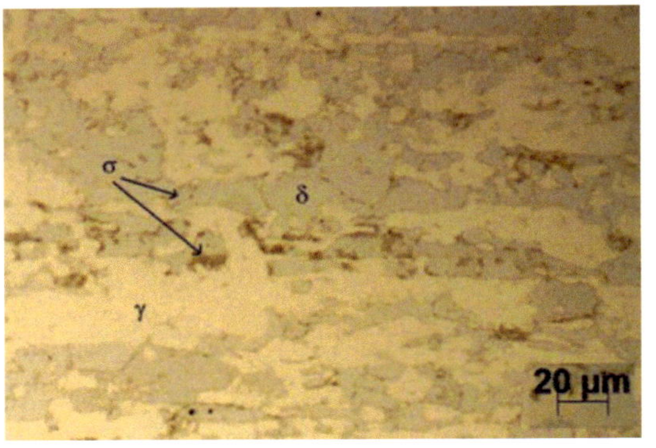

Figure 7. Typical sample after 90 min interval at 800°C.

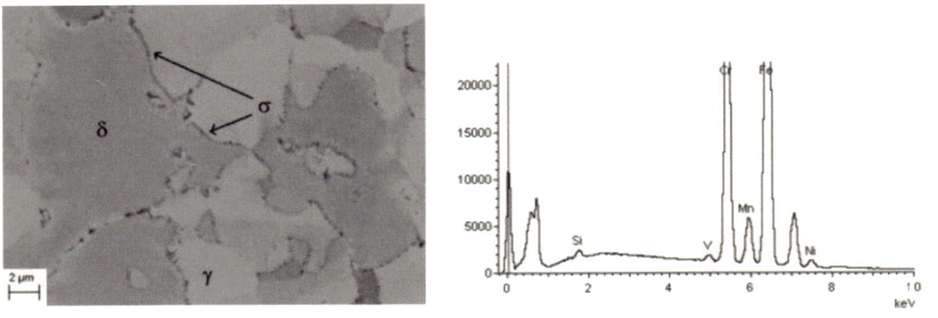

Figure 8. SEM micrographs with the EDS analysis, 30min interval at 800°C.

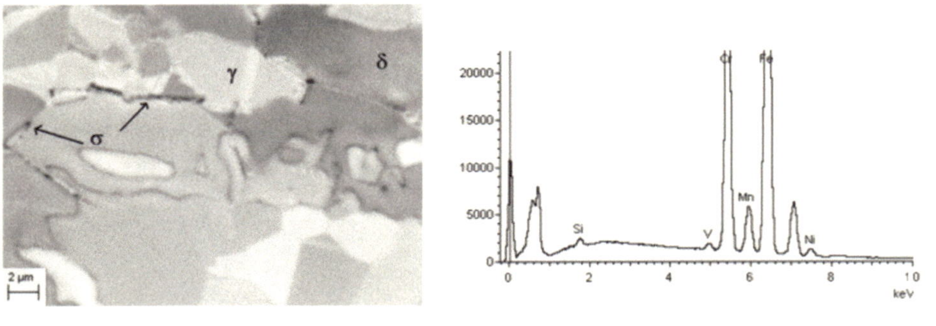

Figure 9. SEM micrographs with the EDS analysis, 60 min interval at 800°C.

Scanning electron microscopy analysis exhibited a similar phase contrast when compared to the optical microscopy results. In this case, three distinct gray areas are observed and related to ferrite (δ), austenite (γ), and sigma phase (σ).

EDS analysis of these same samples observed now by SEM (**Figures 8–10**) exhibits high concentrations of chromium and iron, especially at the grain interfaces, where a darker gray phase is observed. Because sigma phase is mainly formed by Cr and Fe and preferentially has its segregation at the ferrite/ferrite and austenite/ferrite interfaces, one can relate these areas as sigma sites.

http://dx.doi.org/10.5772/intechopen.81128

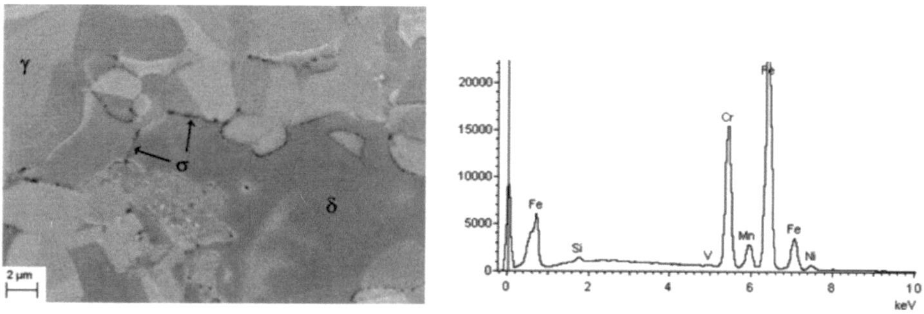

Figure 10. SEM micrographs with the EDS analysis, 90 min interval at 800°C.

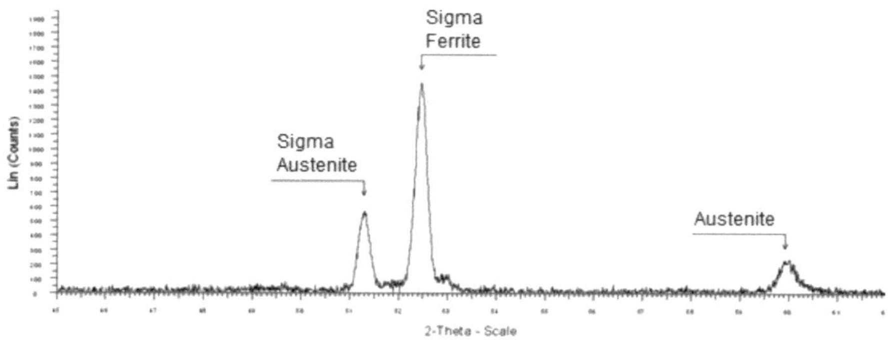

Figure 11. X-ray diffraction pattern at 800°C after 30 min.

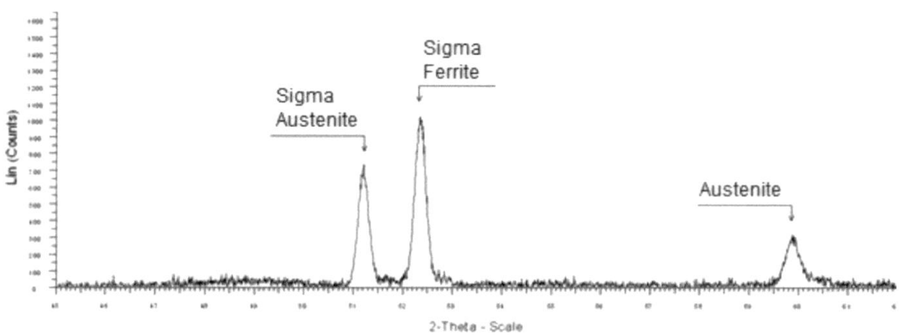

Figure 12. X-ray diffraction pattern at 800°C after 60 min.

5.2. X-ray diffraction patterns

Figures 11–13 show the diffraction patterns of each sample obtained at the end of the dwell time interval (30, 60, and 90 min) at a temperature of 800°C. Through these, it is possible to verify the presence of the ferrite, austenite, and sigma phases.

X-ray diffraction patterns provided the necessary data for the phase quantification in the lean duplex steel. Diffraction was initially performed at room temperature (30°C), followed by

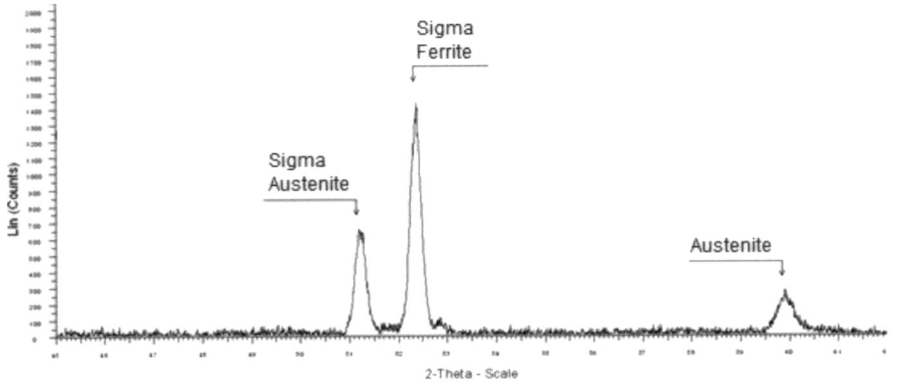

Figure 13. X-ray diffraction pattern at 800°C after 90 min.

Temperature, time	Ferrite	Austenite	Sigma	Ferrite	Austenite	Sigma	Ferrite	Austenite	Sigma
30°C	72.95	27.05	0.00	63.14	36.86	0.00	72.88	27.12	0.00
800°C, 1 min	65.98	32.13	1.89	52.80	44.26	2.94	62.85	36.80	0.35
800°C, 10 min	65.39	31.73	2.88	51.09	42.96	5.95	61.73	37.17	1.10
800°C, 20 min	64.56	32.30	3.14	48.84	43.81	7.35	60.17	37.82	2.01
800°C, 30 min	58.65	29.66	11.49	47.11	43.53	9.36	58.29	38.91	2.80
800°C, 40 min	—	—	—	45.53	43.51	10.96	57.38	39.23	3.39
800°C, 50 min	—	—	—	43.39	43.00	13.61	55.87	38.11	6.03
800°C, 60 min	—	—	—	42.45	41.63	15.92	50.12	34.16	15.72
800°C, 70 min	—	—	—	—	—	—	49.83	34.51	15.67
800°C, 80 min	—	—	—	—	—	—	49.30	34.91	15.79
800°C, 90 min	—	—	—	—	—	—	48.39	34.09	17.52

Table 2. Phase quantification of the lean duplex steel (%vol).

various scans at 800°C at each distinct time interval. **Table 2** presents the values obtained for each phase in the different scan times.

6. The kinetics of phase transformations

In situ X-ray analysis has shown to be effective to determine the kinetics of phase transformation for the evaluation of lean duplex steel phases. **Figures 14–16** present the plotted curves for the transformed amounts of sigma, ferrite, and austenite during the 800°C isothermal heating.

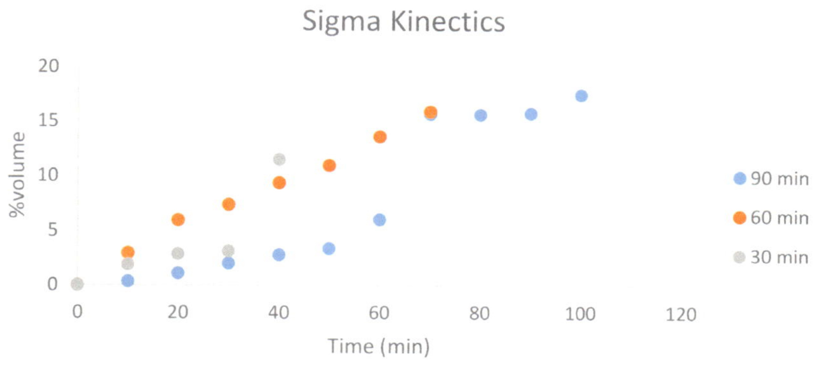

Figure 14. Isothermal evolution of the sigma phase.

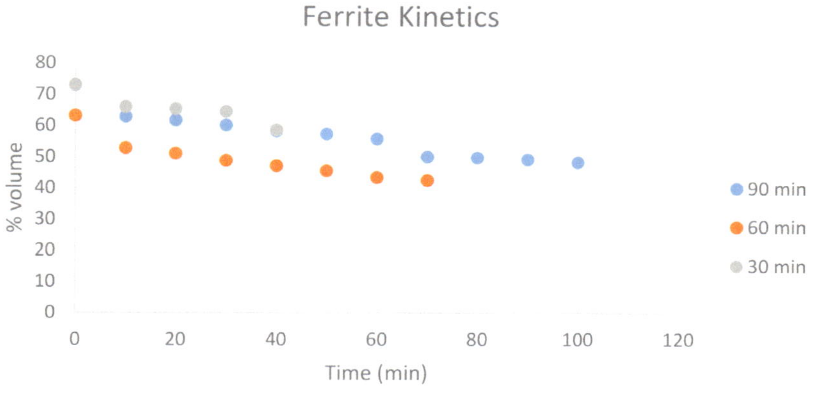

Figure 15. Isothermal evolution of the ferrite phase.

When comparing the evolution of sigma phase (σ) for the 30, 60, and 90 min samples (**Figure 14**), it can be observed that the only one of the three samples that presented a quasi-linear evolution was the 60 min sample, while the two others had a quite variable tendency regarding their volume increase during their isothermal heating. This behavior might be explained by the initial amounts of ferrite and austenite in the 30 and 90 min samples. Both samples had the bigger ferrite/austenite imbalance—that difference in their volume percentage might explain the discontinuities in the speed of sigma formation for those samples, as shown by previous works [13].

Regarding the ferrite evolution, presented in **Figure 15**, it is quite clear that most of its %vol. decrease is related to the sigma formation. Comparing the values for both phases at the same time intervals, it is observed that there is indeed an increase in sigma volume while the ferrite volumes decrease, as reported earlier. That cannot be said from the austenite phase, shown in **Figure 16**. In fact, it is observed that there is a slight increase in austenite volumes along the isothermal heating.

These results lead to the assumption that the sigma formation mechanism is either by continuous or by discontinuous precipitation from ferrite or by a eutectoid decomposition of

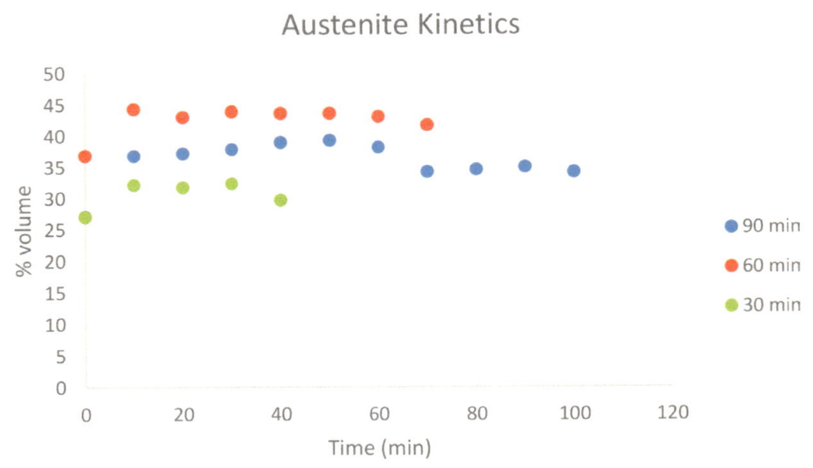

Figure 16. Isothermal evolution of the austenite phase.

ferrite, where a significant decrease in the amounts of ferrite is observed. For this study, it is more likely that the formation of sigma phase is due to the eutectoid decomposition of ferrite because this type of mechanism leads also to the formation of amounts of secondary austenite, therefore increasing its volume over time.

Both mechanisms though lead to Cr- and Mo-depleted phases that will reduce the corrosion resistance of the lean duplex steel, particularly reducing pitting corrosion resistance as reported by earlier works [19, 20]. Wilms et al. [21] also indicated that the formation of sigma has a strong effect on the mechanical properties of stainless steels.

7. Conclusions

Samples of steel UNS S32304 were subjected to isothermal heating at 800°C for 30, 60, and 90 min while analyzed by in situ X-ray diffraction to evaluate the phase transformations occurring at this temperature and time periods. The results lead to the following conclusions:

- Significant amounts of sigma phase were formed during isothermal heating for three time intervals.
- Sigma phase precipitation was observed both in the ferrite/austenite and ferrite/ferrite grain boundaries.
- Depending on the ferrite/austenite balance, sigma phase kinetics may lead to volume percentage of about 15% after 60 min and 17% after a period of 90 min.
- The most likely mechanism for sigma formation on the assessed lean duplex steel is the eutectoid decomposition of ferrite because this type of mechanism leads also to the formation of amounts of secondary austenite, therefore increasing its volume over time.

- Such high percentages of sigma formation lead to Cr- and Mo-depleted phases that will reduce the corrosion resistance of the lean duplex steel, particularly reducing pitting corrosion resistance and jeopardizing the mechanical properties of lean duplex stainless steels.

Acknowledgements

The authors would like to thank CNPq (Brazilian National Council for Scientific and Technological Development) for the support in this research.

Conflict of interest

The authors declare no conflicts of interest.

Author details

Adriana da Cunha Rocha*, Andrea Pedroza da Rocha Santos and Gabriela Ribeiro Pereira

*Address all correspondence to: adrirocha@metalmat.ufrj.br

Department of Metallurgical and Materials Engineering, Federal University of Rio de Janeiro (UFRJ), Rio de Janeiro, RJ, Brazil

References

[1] Davis JR. ASM Specialty Handbook—Stainless Steel. 1st ed. ASM International, Materials Park, Ohio, USA; 1994

[2] Brytan Z, Niagaj J. The lean duplex stainless steel welded joint after isothermal aging heat treatment. Archives of Materials Science and Engineering. 2013;**60**(1):24-31

[3] Souza CS, Lins VFC, Silveira DM. Avaliação da soldagem multipasse de chapas espessas de aços inoxidáveis lean duplex UNS S32304 soldadas pelos processos SMAW, GMAW e FCAW—Parte II: Resistência à corrosão. Soldagem e Inspeção. 2013;**18**(03):257-267

[4] Practical Guidelines for the Fabrication of Duplex Stainless Steels. 2nd ed. Prepared by TMR Stainless. Pittsburgh, PA, USA, London: International Molybdenum Association; 2009

[5] Gunn RN. Duplex Stainless Steels: Microstructures, Properties and Applications. 1st ed. England: Abington Publishing; 1997

[6] Chen TH, Yang JR. Effects of solution treatment and continuous cooling on σ-phase precipitation in a 2205 duplex stainless steel. Materials Science and Engineering A. 2001; **311**(1-2):28-41

[7] Alvarez-Armas I, Degallaix-Moreui S. Duplex Stainless Steels. 1st ed. London: ISTE; 2009

[8] Hall EO, Algie SH. The Sigma Phase. Metallurgical Reviews. 1966;**11**(1):61-88

[9] Calliari I, Brunelli K, Dabala M, et al. Measuring secondary phases in duplex stainless steels. The Journal of the Minerals, Metals & Materials Society. 2009;**61**(1):80-83

[10] Elmer JW, Palmer TA, Specht ED. In situ observations of sigma phase dissolution in 2205 duplex stainless steel using synchrotron X-ray diffraction. Materials Science and Engineering A. 2007;**459**(1-2):151-155

[11] Hsieh C, Wu W. Overview of intermetallic sigma (σ) phase precipitation in stainless steels. ISRN Metallurgy. 2012;**2012**:1-16

[12] Villanueva DME, Junior FCP, Plaut RL, Padilha AF. Comparative study on sigma phase precipitation of three types of stainless steels: Austenitic, superferritic and duplex. Materials Science and Technology. 2006;**22**(9):1098-1104

[13] Magnabosco R. Kinetics of sigma phase formation in a duplex stainless steel. Materials Research. 2009;**12**(3):321-327

[14] Pohl M, Storz O, Glogowski T. Effect of intermetallic precipitations on the properties of duplex stainless steel. Materials Characterization. 2007;**58**(1):65-71

[15] Cullity BD. Elements of X-ray Diffraction. 1st ed. Massachusetts, USA: Addison-Wesley Publishing Company, Reading; 1978

[16] Young RA. The Rietveld Method. 1st ed. New York: Oxford University Press; 1995

[17] Rietveld HM. A profile refinement method for nuclear and magnetic structures. Journal of Applied Crystallography. 1969;**2**(2):65-71

[18] Jackson EMLEM, Visser PE, Cornish LA. Distinguishing between chi and sigma phases in duplex stainless steels using potentiostatic etching. Materials Characterization. 1993;**31**(4):185-190

[19] Fang YL, Liu ZY, Xue WY, et al. Precipitation of secondary phases in lean duplex stainless steel 2101 during isothermal ageing. ISIJ International. 2010;**50**(2):286-293

[20] Magnabosco R, Alonso-Falleiros N. Pit morphology and its relation to microstructure of 850°C aged duplex stainless steel, Corrosion. 2005;**61**:130-136

[21] Wilms ME, Gadgil VJ, Krougman JM, Kolster BH. The effect of σ-phase precipitation at 800°C on the mechanical properties of a high alloyed duplex stainless steel. Materials at High Temperatures. 1991;**9**(3):160-166

Chapter 2

Machining of Stainless Steels and Alloys Using Non-Traditional Machining Processes

Ramesh Rudrapati

Additional information is available at the end of the chapter

http://dx.doi.org/10.5772/intechopen.81611

Abstract

Stainless steels and alloys are characterized primarily by their corrosion resistance, high strength, ductility, etc. used for various advanced applications like automotive and aerospace, sugar refineries, construction materials, etc. Many advanced high-speed machineries /systems need fine quality of parts to provide good performance in its working conditions. The machining of stainless steel and its alloys is of interest, because, of its excellent mechanical properties. Stainless steels and alloys are machined generally by traditional machining processes. But complex shapes and features on products are difficult task with the use of traditional metal cutting techniques. To machine the advanced materials to produce high dimensional accuracy and generation of intricate shapes in difficult-to-machine materials like stainless steels and alloys, nontraditional machining (NTM) techniques are now attractive the viable choices. To attain improved machining performance of the NTM processes, it is always necessary to find the optimal combinations of various process input parameters of those processes. In the present chapter, some aspects of machining of stainless steel and alloys using NTM processes such as electric discharge machining (EDM) and wire EDM, are discussed and some concluding remarks have been drawn from the study.

Keywords: stainless steels and alloys, nontraditional machining, EDM, wire EDM

1. Introduction

With the increased industrial and technological growth, materials exhibiting excellent strength, hardness, stiffness to strength ratio, good corrosion resistance properties have found wide applications particularly in aerospace, nuclear engineering, navel, marine, space technology, tool industries [1], etc. The advanced materials include stainless steel, refractory materials and its alloys, carbides, ceramic alloys, glass and fiber-based composites, etc. Stainless steel and its alloys got immense importance for variety of advanced applications. Most of stainless steels and its alloys (SS&A) treated as difficult to machine materials. Machining of SS&A using traditional methods like turning, milling, drilling, grinding, etc., cannot preferable, because of low cutting speeds and material removal rates. And it's difficult to find cutting tools to cut these type materials. Machining of stainless steels and its alloys is an important research area for industrialists.

Stainless steel materials gained much important materials for industries at the beginning of twentieth century. Stainless steels are defined as steel alloys described mainly by its corrosion resistance, high strength, high ductility and high chromium content present.

- Steel is one of the most attractive and widely used products in the world for various applications.
- Currently, the steel industry is plays vital role for process of change.
- Because of continuing procedural, technical and economic growth, the manufacture and use of steel is becoming increasing
- The use of steel continuously gaining its share of world-wide over the last three decades

French scientist Berthier was found the iron-chromium alloys which resist to certain acids in early eighteenth century. Then after Brustline another French scientist identified the importance of carbon (C) percentage levels and chromium (Cr) levels. In 1904, Leon Guilet [2] had invented the martensitic and ferritic SS-alloys series (i.e. 410, 442, 446, and 440 C) and Fe-Ni-Cr austenitic alloy series (300 series of SS). Various researchers conducted research investigations to improve the mechanical properties of stainless steels and its alloys with addition of different supplements like Cr, Ni, C, Mo, manganese, silicon, titanium, molybdenum, aluminum, sulfur, phosphorous, nitrogen, etc., to iron (Fe). Monnart and Eduard Maurer invented the stainless steel and austenitic stainless steel in the year of 1911 and 1912 respectively [3], with improved corrosion resistance property. The free machining stainless steels with addition of 15% sulfur was introduced to the world in the year of 1928. Stainless steels with 15% sulfur were called as martensitic 416 stainless. Then after sulfur and phosphorous were both added to make the austenitic stainless grades of 303 [4]. Maurer and Brearly described the advantages of stainless steel for industrial applications [5].

These alloys are milled into coils, sheets, plates, bars, wire, tubing, etc., to be used extensively in cookware, cutlery, surgical appliances, industrial equipment (e.g., in sugar refineries), automotive and aerospace operational alloys, as a construction material, etc. stainless steels and alloys are widely used in storage tanks for transporting of various liquid based materials like orange juice, apple juice and other oil-based products, due to its corrosion resistance and antibacterial properties. These materials are also used in commercial kitchens, food processing plants etc.

The applications of stainless steels and its alloys have been increased enormously over last 30 years in various engineering related fields because of their excellent materials properties such as high strength, high temperature strength, high corrosion resistance, oxidation resistance, etc. These materials are treated as difficulty to machine materials because of low thermal conductivity, high coefficient of thermal expansion, high ductility, and high work-hardening rate, etc. Low thermal conductivity nature in these materials increases in tool temperature during

http://dx.doi.org/10.5772/intechopen.81611

machining, which leads to reduced tool life and tolerances of work-piece. Chip formation is severely affected due to high work-hardening rate and low thermal conductivity. High ductility nature has creates built-up edge on cutting tool edge and propagates vibration and chatter finally it have detrimental effects on quality of machined surface.

It is difficult to find adequate cutting tool materials to machine the advanced materials like stainless steel and its alloys using traditional machining processes. The new cutting tools along with advanced machining methodology called non-traditional machining (NTM) have been developed to machine the newly developed materials like stainless steels, ceramics, high strength polymers, composites and other alloy materials. These processes can capable of machine a wide spectrum of difficult-to-cut materials irrespective of their hardness. The important NTM processes are electric discharge machining (EDM), wire EDM, electron beam machining (EBM), ion beam machining (IBM), etc. It is necessary to conduct the NTM processes in optimum manner to enhance machining performance.

In the present paper reviews the research work carried out from the spin-off from the NTM processes to the development of NTM processes related to thermal energy-based techniques like EDM and WEDM. It reports on the research work involving the optimization of process parameters which expected to influencing the machining performance and productivity. The paper also highlights the important process parameters for performing the above said processes when machining of stainless steels and its alloys.

2. Machining of stainless steel and alloys by NTM processes

Machining or metal cutting is one of the important manufacturing processes in which excess or unwanted material is removed by cutting tool / electrode tool with or without physical contact with the work-piece. The classification of the machining processes is shown in **Figure 1**. From **Figure 1**, one notified that machining process is categorized into two types: traditional machining (TM) processes and non-traditional or non-conventional machining (NTM) processes. In traditional machining processes, material removal has carried out by using sharp edge cutting tool with physical contact with the work material. Tool material wear, generation of vibrations, difficulty of machining hard materials, impossible or uneconomical to produce complex shapes on parts, minimized tool life, decreased production rate and increased production costs are some of the limitations associated in TM processes because of physical contact with the work-piece. To overcome the limitations of TM techniques, NTM are developed. NTM methods applied successfully where work-piece surface is too hard, strong, flexible to resist cutting forces, difficult to clamp, to create complex shapes with both internal and external profiles, to produce fine surface finish and tolerances, residual stresses are unacceptable for materials.

In NTM processes, physical tool may not present, no chip formation may take place, cutting tool need not to be harder than work material and it do not necessarily apply mechanical energy. NTM methods are classified according to usage of energy for machining are chemical energy, electrical energy, mechanical energy, electrical energy, thermal energy, etc. Thermal energy-based processes are plays important role in metal cutting industries. In thermal energy-based techniques, material

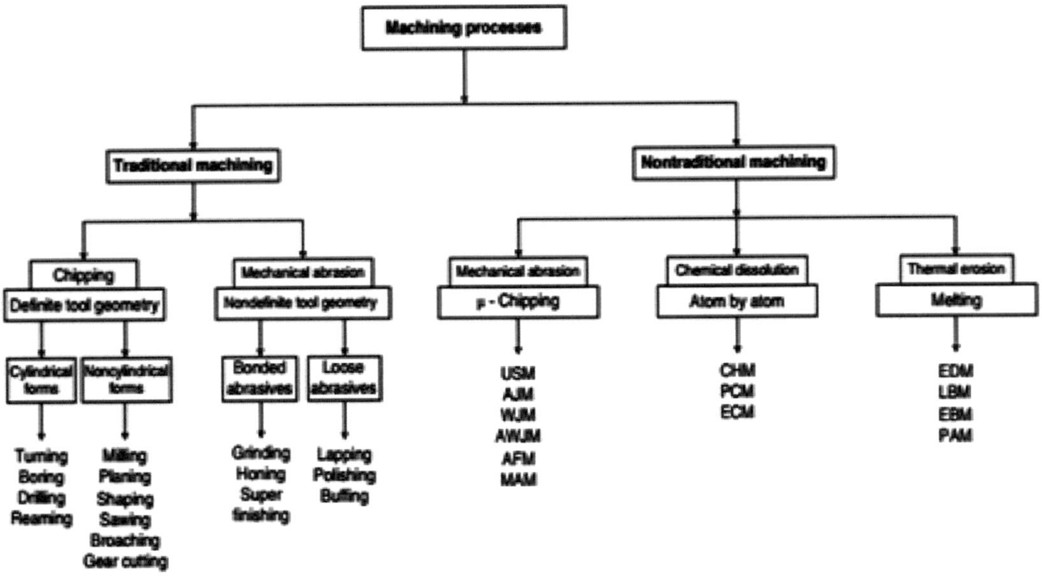

Figure 1. Classification of machining processes [2].

removal processes based on thermal energy are characterized by very high local temperatures—hot enough to remove excess material by fusion or vaporization method. Because of the high temperature generation during metal removal process, these processes cause physical and metallurgical damage to the newly generated surface of the work material. The two main processes in this category are (1) electric discharge machining and (2) wire electric discharge machining. These processes can be used only on electrically conducting work materials.

2.1. Electric discharge machining (EDM)

Electric discharge machining (EDM) is one of the nontraditional processes most widely used in mold, die, aerospace, automotive industry and surgical components for generating intricately shaped, and mold cavities on difficult-to-cut electrically conductive materials like high strength, temperature resistant (HSTR) alloys [6–10]. In EDM, mechanical stresses, chatter and vibration during machining can be eliminated because electrode tool does not make any direct contact with the work-piece [9].

An EDM setup is illustrated in **Figure 2**. The shape of the finished work-piece surface is replica of formed electrode tool. The EDM process conducted in the presence of a dielectric fluid, which makes a path for passing each discharge current to the fluid. So, fluid becomes ionized in the gap. The discharges are generated by a pulsating direct current and power supply connected to the work and the tool. The sparks occur across a small gap between electrode tool and work-piece surface. There are hundreds of electrical discharges generated during machining process and every discharge energy may form a specific surface integrity on the part which results in creation of surface texture by overlapping craters [11].

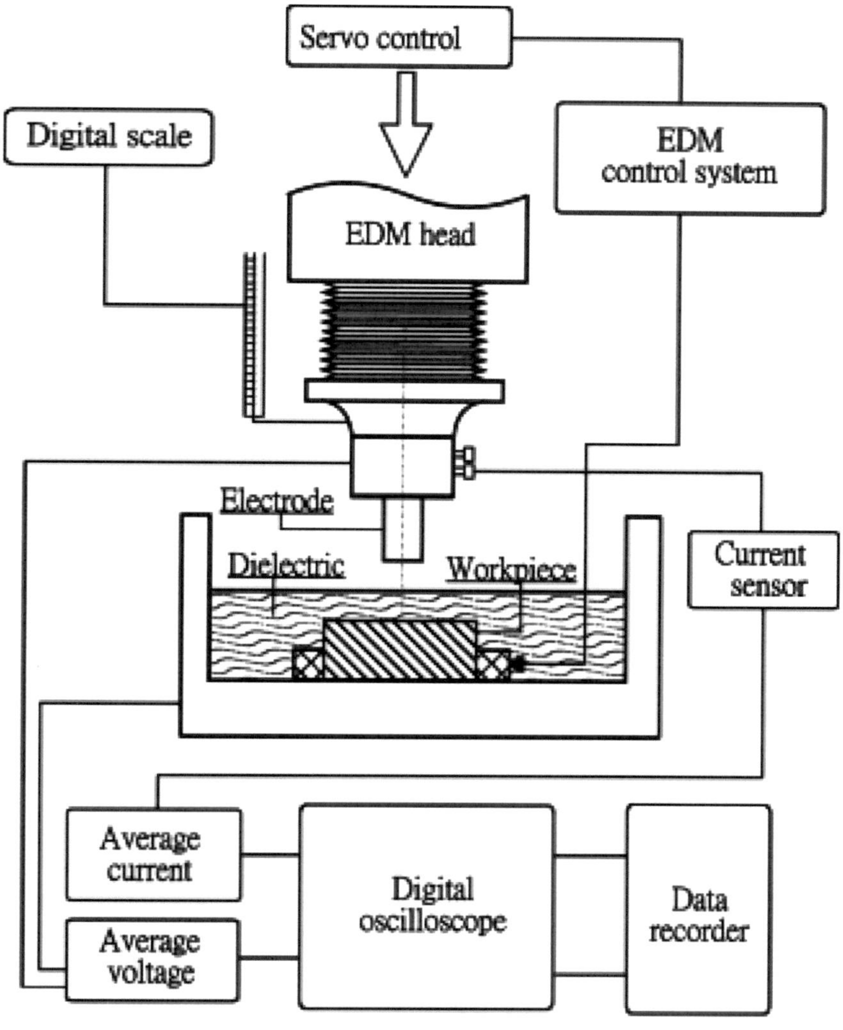

Figure 2. EDM process [20].

The mechanism behind the material removal in EDM is complex and difficult to understand and thus it is difficult task to formulate the relationships between the input and output parameters and to predict the better EDM performance [9, 12, 13]. The inappropriate selection of control parameters in EDM process may result in poor machining performance, short-circuiting between the electrodes or even work-piece surface may damage thereby reducing the productivity and affecting the quality. Due to high capital and manufacturing costs, these processes need to be operate economically and predictably to obtain machining economics. Identifying optimal set of parametric conditions is necessary to optimize production rate, machining time, lower material waste and costs, [8, 14] etc. Analyzing, modeling and optimizing EDM process may useful to improve the performance characteristics of EDMed parts.

Literature survey has been done to review and study the machining of stainless steels and its alloys in EDM process. The details of the literature survey are given as follows:

The maraging steel is a high strength material which exhibits resistance to corrosion oxidation in its working. Machining of maraging steels by using traditional machining process is difficult, because, of its high hardness. Ruma et al. [1] had experimented the EDM of maraging steel material to investigate the influences of EDM parameters on output responses. Researchers optimized the multi-performance characteristics using teaching learning-based optimization (TLBO) and genetic algorithm (GA). Authors mentioned that selection of optimum parametric conditions essential for conduction of EDM process effectively and economically. The 17-4 PH steels widely used for aerospace and die manufacturing, Chandramouli and Eswaraiah [6] had experimented EDM of 17-4 precipitation hardening stainless steel material to evaluate the influences of input parameters: peak current, pulse on time (T_{ONN}), pulse off time (T_{OFF}) and tool lift time on output response: material removal rate (MRR) and surface roughness. Researchers had studied, analyzed and optimized the EDM process to fine quality parameter (R_a) and qualitative factor, MRR when machining of 17-4 PH steels. Die steel materials are most widely used for industrial purposes and machining of these materials are important area of research. Banh et al. [14] presented a research work related to optimize the quality characteristics of die steel materials such as SKD61, SKD11 and SKT4, in titanium powder mixed EDM process. Researchers had been considered workpiece material, tool material, polarity, T_{ONN}, current, T_{OFF}, and powder concentration as input parameters and MRR, tool wear rate, surface roughness, and microhardness surface, as quality responses. They stated that input parameter selection is very crucial for obtaining better performance characteristics. Electric discharge machining with powder mixed has improved its performance significantly. Priyaranjan et al. [15] had made an experimental analysis to evaluate the relationship between process parameters like pulse current, pulse duration, T_{OFF} and dielectric pressure, and performance characteristics of MRR, electrode wear rate and taper angle of the hole of AISI 329 stainless steel in EDM process. The stainless-steel grade of 329 which commonly found for several industrial applications like heat exchangers, chemical tanks, pressure vessels, etc. Investigators had been effectively optimized output responses by controlling input parameters. They stated from their work that properties of electrode and work-piece material influences much for EDM when machining of AISI 329 stainless steel.

Pradhan and Das [7] had made experimental analysis and developed predictive models for optimizing MRR by considering discharge current, pulse duration, duty cycle, and voltage as input machining parameters in EDM of AISI D2 tool steel. They identified from their study that MRR of D2 tool steel in EDM is highly effected by input parameters. The MRR is increase with increase of discharge current, duty cycle, and voltage whereas pulse duration is most significant for MRR. Janmanee and Muttamara [16] have performed the experiment on EDM to improve the drilling performance of stainless steel AISI 431 using brass tube electrode. They determined that MRR increases with increase of servo rate, while the taper of the hole increases with increase of electrical current and servo rate. Panda and Bhoi [10] worked on analyzing, modeling and optimization of MRR for die steel in EDM process using ANN. They considered pulse current, pulse voltage and duty factor as input parameters. Researchers had been developed relationships between input parameters and MRR effectively. AISI P20 steel is using as a tool for injection molding, Amorim and Weingaertner [17] had made an

http://dx.doi.org/10.5772/intechopen.81611

experimental research work based on the EDM of AISI P20 tool steel. They analyzed the significances of control parameters on MRR, volumetric relative wear (VRW) and surface roughness of P20 tool steel material. Kumagai et al. [18] had explored the effects of electrode material on narrow deep holes in drilling of steel blocks. Jain [19] carried out the experimental analysis to identify the effects of pulse time, diameter of tool and depth of penetration on output responses of EDM of high-speed steel (HSS). He mentioned from his study that EDM is very efficient to make blind holes in HSS material. Some more publications related to the EDM of stainless steels and alloys are given in **Table 1**.

2.2. WEDM

Wire electrical discharge machining (WEDM) is a non-traditional machining process that is widely used to machine high strength materials. It is one form of EDM as mentioned earlier.

S. No	Reference	Work material	Process parameters	Output responses	Impact
1	Ubaid et al. [21]	Stainless steel 304	Current, T_{OFF}, T_{ON}	MRR, EWR, VEW	Current and T_{OFF} are most significant
2	Majumder [22]	AISI 316 LN stainless steel	Current, T_{OFF}, T_{ON}	MRR, EWR	All parameters are significant
3	Priyaranjan et al. [15]	AISI 329 stainless steel	Current, T_{OFF}, pulse duration, dielectric pressure	MRR, EWR, taper angle of hole	Current and pulse duration are significant for brass electrode; T_{OFF} is significant for copper electrode
4	Dastagiri and Kumar [23]	Stainless steel and En41b steel	Current, V, T_{ON} and duty factor	MRR, R_a and hardness	All the parameters are significant for both the materials
5	Shashikant et al. [24]	EN41 steel	T_{ON}, T_{OFF}, V and discharge current	R_a, Rq, Rsk, Rku and Rsm	Current is significant for surface roughness
6	Shashikant et al. [25]	EN19 and EN41	T_{OFF}, T_{ON}, V and discharge current	MRR	Current is significant for MRR of both the materials
7	Rajmohan et al. [26]	Stainless steel 304	Current and voltage, T_{OFF}, T_{ON}	MRR	Current and T_{OFF} are significant for MRR
8	Faisal and Kumar [27]	EN 31 steel	Current, T_{ON}, T_{OFF}, gap voltage	MRR and R_a	Current is significant for both the responses
9	Behera et al. [28]	AISI 304 stainless steel	Current, T_{ON}, flushing pressure	MRR	Current is significant for MRR
10	Kumar et al. [29]	AISI H13 die steel	Sparking voltage, discharge current, T_{OFF}, T_{ON}, dielectric fluid, workpiece material H13 hot die steel: hardened and tempered	Micro-hardness, R_a	All parameters have considerable effects on responses

T_{OFF} = pulse off time; T_{ON} = pulse on time; EWR = electrode wear rate; VEW = volumetric electrode wear; MRR = material removal rate; R_a = surface roughness.

Table 1. Literature survey related to machining of stainless steels and alloys in EDM.

WEDM is very complex machining process which used to create high complicated shapes such as tapers, involutes, parabolas and ellipses [30] on hard materials like stainless steel, nitronic austenitic stainless, beryllium copper and titanium, ceramics, metal matrix composites [31], etc. The WEDM process involves a material erosion mechanism by pulsing of direct current between the wire electrode and work piece [32]. WEDM uses electro- thermal based energy to cut electrically conductive material(s). WEDM uses a thin continuously moving electrode in the form of wire in the range of 0.050–0.35 mm [33]. In WEDM, unwanted material is removed using series of discrete discharge currents between the wire electrode and the work material in the presence of a dielectric fluid [34]. Each discharge passed to the dielectric fluid and it becomes ionized in the gap resulted in generation of extremely high temperatures, therefore work-piece surface is melted and removed. The machined surface is cleaned by dielectric fluid by taking removed chip along with it. The schematic diagram of WEDM is shown in **Figure 3**. The melting point temperature of the work material is very important factor for WEDM process rather than its strength and hardness of the material [35].

WEDM is special from of EDM which enables to create very complex shapes very easily by using thin wire as electrode as mentioned above. Here, literature survey has been made to review and investigate the parts produced by WEDM of stainless steels and alloys, those details are given as follows:

DC53 is cold die steel improved from familiar cold die steel SKD1, made from daido steel. Machining of DC53 die steel is an important task. Kanlayasiri and Boonmung [35] presented an investigation of the effects of machining variables: pulse-peak current, T_{ONN}, T_{OFF}, and wire tension on the surface roughness of DC53 die steel material in wire-EDM process. Shahali

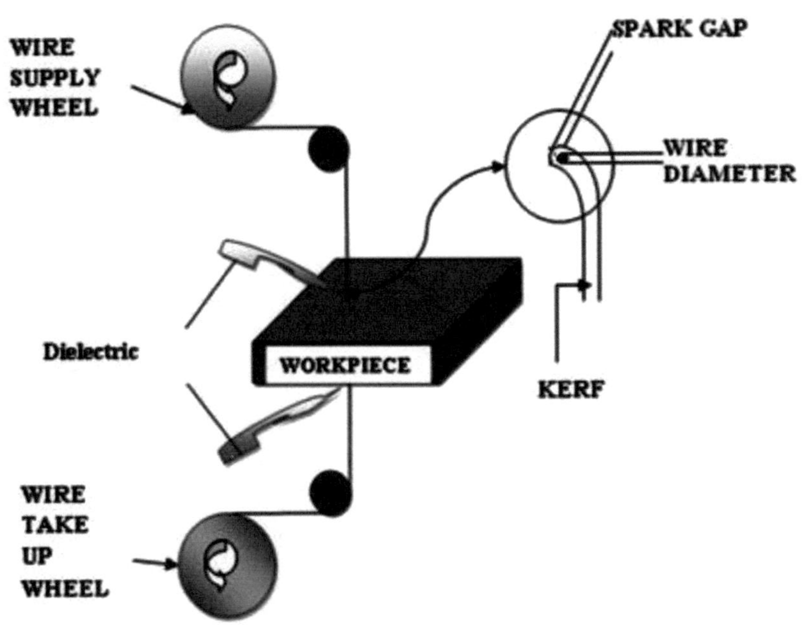

Figure 3. WEDM process [33].

S. No	Reference	Work material	Process parameters	Output responses	Impact
1	Majumder et al. [41]	Indian reduced activation ferritic martensitic (RAFM) steel	pulse current, T_{OFF}, T_{ON}, wire tension	Cutting speed and R_a	All parameters have significant effect on both the responses
2	Giduturi and kumar [42]	H-13 tool steel	Wire tension, wire feed, T_{OFF}, T_{ON}, servo voltage, peak current	MRR and R_a	All parameters have significant effect on both the responses
3	Pamnani et al. [43]	DMR249A steel	Current, torch speed, arc gap	Depth of penetration	Current and torch speed have significant effect on depth of penetration
4	Choudhuri et al. [44]	AISI stainless steel-316	Servo gap set voltage, T_{OFF}, T_{ON} and wire feed rate	MRR, R_a and lower wire consumption	Servo gap set voltage is significant for all the responses
5	Sudhakara and Prasanthi [45]	Powder metallurgical cold worked tool steel	T_{ON}, T_{OFF}, Peak current, spark gap set voltage, wire tension, water pressure	R_a	T_{ON} parameters has the significant effect on R_a
6	Reddy et al. [46]	P20 die-tool steel	Bed speed, T_{ON}, T_{OFF} and current	R_a and volumetric material removal rate	All parameters have significant effect on both the responses
7	Manjaiah et al. [47]	D2 steel	Servo voltage, wire feed, T_{ON}, T_{OFF},	MRR and R_a	T_{ON} and servo voltage have significant effect for both the responses
8	Tosun et al. [48]	AISI 4140 steel	pulse duration, open circuit voltage, wire speed dielectric flushing pressure	R_a	open circuit voltage and pulse duration have significant effect on R_a
9	Ugrasen et al. [49]	Modified AISI 420 steel	T_{ON}, T_{OFF}, current, bed speed	R_a, VMRR, accuracy and EWR	All factors significantly affecting the all responses
10	Sudhakara and Prasanthi [50]	VANADIS 4E (powder metallurgical cold worked tool steel)	T_{ON}, T_{OFF}, peak current, spark gap set voltage, wire tension, water pressure	Dimensional deviation	T_{ON}, T_{OFF}, peak current, water pressure have significant effect on dimensional deviation

VMRR = volumetric material removal rate; T_{OFF} = pulse off time; T_{ON} = pulse on time; EWR = electrode wear rate; VEW = volumetric electrode wear; MRR = material removal rate; R_a = surface roughness.

Table 2. Literature survey related to machining of stainless steels and alloys in WEDM.

et al. [36] made an experimental analysis to identify the optimum input parameters in WEDM of DIN 1.4542 hardening stainless-steel alloy. They considered power, T_{OFF}, voltage and servo and number of finish passes, as input parameters and surface roughness and maximum thickness of white layer as output responses. Hassan et al. [30] presented the experimental work to study the effects of process parameters: pulsed current and pulse-on duration on surface texture of AISI 4140 steel in WEDM process. They found from their study that pulse-on duration is most influential factor than pulsed current to defining the WEDM surface texture.

The P91 steel is extensively used for energy-based industries such as reactor components due to its excellent mechanical properties like high creep strength and thermal conductivity, low thermal expansion, good corrosion resistance, [37, 38] etc. Bhattacharya et al. [32] studied the corrosion behavior of P91 steel material in WEDM process. They stated that WEDMed surfaces have fine passivation than the diamond polished surface.

AISI D3 die-steel possess high resistance to wear and exhibits resistance to heavy pressure which extensively used for applications like blanking, stamping and cold forming dies and punches for long runs and lamination dies [39]. Muthukumar et al. [39] conducted experiments to study the influences of input factors such as T_{ON}, T_{OFF}, gap voltage, wire feed on MRR, R_a and kerf width in WEDM of AISI D3 die-steel. They found improved quality responses through systematic analysis of WEDMed die steels.

SKD11 steel is a high-carbon, high-chromium alloy tool steel which used to make dies, precision gauges, spindle, jigs and fixtures, [40] etc. Zhang et al. [40] had made research analysis to seek optimal output responses: MRR and 3D surface quality (Sq) by controlling the input parameters in WEDM of SKD11 steel. Investigators had been improved the performance of WEDM process by systematic analysis and optimization.

One can found more reported articles on machining of stainless steels and alloys in WEDM from the literature. Here, some of the published papers have been presented in **Table 2.**

3. Conclusions

Followings are the conclusions drawn from the present study of machining of stainless steel and its alloys using EDM and WEDM:

1. A historical review of steel and its alloys have been discussed.
2. Stainless steels and alloys are the widely used industrial material for various advanced applications like aerospace, automobile, surgical, mold, tool and die industries
3. The importance of machining of stainless steels and alloys have been discussed
4. NTM processes are extensively used for producing complicated shapes or features on these materials
5. EDM and WEDM are the important NTM methods which highly used by metal cutting
6. Selection of input process parameters is very important for both EDM and WEDM techniques
7. The effects of input parameters on output performance characteristics of EDMed or WEDMed parts needs to be understand properly

8. Systematic analysis and proper understanding of these methods are important to make effective use of EDM and WEDM processes

9. More extensive research needs to be conducted on stainless steel and alloys in EDM and WEDM methods to enhance the performance of machining

Acknowledgements

The author giving thanks to Dr. Eng. Fisiha Getachew, Vice President for Academic Affairs, Hawasaa University (HU), Dr. Migbar Assefa, Assistant Professor & Dean, Faculty of Manufacturing, HU and Mr. Lijalem Mulugeta, Lecturer & HOD, Mechanical Engineering Department, HU - Institute of Technology (IOT) for encouraging me to write this chapter.

Conflict of interest

The author declaring no conflict of interest.

Notes/thanks/other declarations

The author giving thanks to Editor of this book for giving opportunity to write this chapter.

Author details

Ramesh Rudrapati

Address all correspondence to: rameshrudrapati@gmail.com

Department of Mechanical Engineering, Hawassa University, Institute of Technology, Hawassa, Ethiopia

References

[1] Ruma S, Bikash C, Barma JD, Prasun C. Optimization of wire EDM parameters using teaching learning-based algorithm during machining of maraging steel 300. Materials Today: Proceedings. 2018;**5**:7541-7551

[2] Helmi AY. Machining of Stainless Steels and Super Alloys Traditional and Nontraditional Techniques. UK: John Wiley & Sons, Ltd; 2016

[3] John CT. Stainless Steel in the Brewery. MBAA TQ. 2001:67-82

[4] Palmer FR. Ferrous alloys. US Patent: 1961.777

[5] Paar JG, Hansen A. Introduction to Stainless Steel. USA: American Society for Metals; 1965

[6] Chandramouli S, Eswaraiah K. Optimization of EDM process parameters in machining of 17-4 PH steel using Taguchi method. Materials Today Proceedings. 2017;**4**(2):2040-2047

[7] Pradhan MK, Das R. Recurrent neural network estimation of material removal rate in electrical discharge machining of AISI D2 tool steel. Proceedings of the Institution of Mechanical Engineers, Part B: Journal of Engineering Manufacture. 2011;**225**:414-421

[8] Mandal D, Pal SK, Saha P. Modelling of electrical discharge machining process using back propagation neural network and multi-objective optimization using non-dominating sorting genetic algorithm-II. Journal of Material Processing Technology. 2007;**186**(13):154-162

[9] Ho KH, Newman ST. State of the art electrical discharge machining (EDM). International Journal of Machine Tools & Manufacture. 2003;**43**:1287-1300

[10] Panda DK, Bhoi RK. Artificial neural network prediction of material removal rate in electro discharge machining. Materials and Manufacturing Processes. 2005;**20**:645-672

[11] Świercz R, Świercz DO. Experimental investigation of surface layer properties of high thermal conductivity tool steel after electrical discharge machining. Metals. 2017;**7**(12):550. DOI: 10.3390/met7120550

[12] Gostimirovic M, Kovac P, Sekulic M, Skoric B. Influence of discharge energy on machining characteristics in EDM. Journal of Mechanical Science and Technology. 2012;**26**:173-179

[13] Świercz R, Dorota O, Świercz. Influence of electrical discharge pulse energy on the surface integrity of tool steel 1.2713. In: Proceedings of the 26th International Conference on Metallurgy and Materials; Brno, Czech Republic; 24-26 May 2017. 2017

[14] Banh T-L, Nguyen H-P, Ngo C, Nguyen D-T. Characteristics optimization of powder mixed electric discharge machining using titanium powder for die steel materials. Proceedings of the Institution of Mechanical Engineers, Part E: Journal of Engineering Manufacture. 2018;**232**(3):281-298

[15] Priyaranjan S, Sujit S, Dhananjay RM. Electrical discharge machining of AISI 329 stainless steel using copper and brass rotary tubular electrode. Procedia Materials Science. 2014;**5**:1771-1780

[16] Janmanee P, Muttamara A. A Study of Hole Drilling on stainless steel AISI 431 by EDM using brass tube electrode. International Transaction Journal of Engineering, Management, & Applied Science & Technologies. 2011;**2**:471-481

[17] Amorim FL, Weingaertner WL. The influence of generator actuation mode and process parameters on the performance of finish EDM of a tool steel. Journal of Materials Processing Technology. 2005;**166**:411-416

[18] Kumagai S, Misawa N, Takeda K, Abdukarimov ET. Plasma-applied machining of a narrow and deep hole in a metal using a dielectric-encased wire electrode. Thin Solid Films. 2004;**457**:180-185

[19] Jain VK. Analysis of electrical discharge drilling of a precision blind hole in HSS using bit type of tool. Microtecnic. 1989;**2**:34-40

[20] Lin YC, Cheng CH, Su BL, Hwang LR. Machining characteristics and optimization of machining parameters of SKH 57 high-speed steel using electrical discharge machining based on Taguchi method. Materials and Manufacturing Processes. 2006;**21**(8):922-929

[21] Ubaid MA, Fikri TD, Shukry HA, Laith Al-Juboori A. Optimization of electro discharge machining process parameters with fuzzy logic for stainless steel 304 (ASTM A240). Journal of Manufacturing Science and Engineering. 2017;**140**:1-13

[22] Majumder A. Process parameter optimization during EDM of AISI 316 LN stainless steel by using fuzzy based multiobjective PSO. Journal of Mechanical Science and Technology. 2013;**27**:2143-2151

[23] Dastagiri M, Kumar AH. Experimental investigation of EDM parameters on stainless steel & En41b. Procedia Engineering. 2014;**97**:1551-1564

[24] Shashikant V, Roy AK, Kumar K. Effect and optimization of various machine process parameters on the surface roughness in EDM for an EN41 material using grey-Taguchi. Procedia Material Science. 2014;**5**:1702-1709

[25] Shashikant V, Roy AK, Kaushik K. Effect and optimization of machine process parameters on MRR for EN19 & EN41 materials using Taguchi. Procedia Technology. 2014;**14**:204-210

[26] Rajmohan T, Prabhu R, Rao GS, Palanikumar K. Optimization of machining parameters in electrical discharge machining (EDM) of 304 stainless steel. Procedia Engineering. 2012;**38**:1030-1036

[27] Faisal N, Kumar K. Optimization of machine process parameters in EDM for EN 31 using evolutionary optimization technique. Technologies. 2018;**6**(2):54. DOI: 10.3390/technologies6020054

[28] Behera A, Debashrita P, Mayank B. Optimization of process parameters in electro discharge machining of AISI 304 stainless steel. IOP Conference Series: Materials Science and Engineering. 2018;**390**:1-6

[29] Kumar S, Singh R, Singh TP, Sethi BL. Comparison of material transfer in electrical discharge machining of AISI H13 die steel. Proceedings of the Institution of Mechanical Engineers, Part C: Journal of Mechanical Engineering Science. 2009;**223**:1733-1740

[30] Hassan MA, Mehat NS, Sharif S, Daud R, Tomadi SH, Reza MS. IAENG. Study of the surface integrity of AISI 4140 steel in wire electrical discharge machining. In: Proceedings of the International Multi Conference of Engineers and Computer Scientists; Hong Kong. 2009

[31] Ho KH, Newman ST, Rahimifard S, Allen RD. State of the art in wire electrical discharge machining (WEDM). International Journal of Machine Tools & Manufacture. 2004;**44**:1247-1259

[32] Bhattacharya S, Geogy JA, Abhishek M, Vivekanand K, Dey GK. Corrosion behavior of wire electrical discharge machined surfaces of P91 steel. Journal of Materials Engineering and Performance. 2018;**27**:4561-4570

[33] Kapoor J, Singh S, Jaimal SK. High-performance wire electrodes for wire electrical-discharge machining—A review. Proceedings of the Institution of Mechanical Engineers, Part B: Journal of Engineering Manufacture. 2012;**226**:1757-1773

[34] Benedict GF. Nontraditional Machining Processes. New York & Basel: Marcel Dekker Inc; 1987. pp. 231-232

[35] Kanlayasiri K, Boonmung S. Effects of wire-EDM machining variables on surface roughness of newly developed DC 53 die steel: Design of experiments and regression model. Journal of Materials Processing Technology. 2004;**192-193**:459-464

[36] Shahali H, Yazdi MRS, Aminollah M, Ehsan I. Optimization of surface roughness and thickness of white layer in wire electrical discharge machining of DIN 1.4542 stainless steel using micro-genetic algorithm and signal to noise ratio techniques. Proceedings of the Institution of Mechanical Engineers, Part B: Journal of Engineering Manufacture. 2012;**226**:803-812

[37] Tavassoli AAF, Diegele E, Lindau R, Luzginova N, Tanigawa H. Current status and recent research achievements in ferritic/martensitic steels. Journal of Nuclear Materials. 2014;**455**:269-276

[38] Shamardin V, Golovanov V, Bulanova T, Povstianko A, Fedoseev AE, Goncharenko YA, et al. Mechanical properties and microstructure of advanced ferritic-martensitic steels used under high dose neutron irradiation. Journal of Nuclear Materials. 1999;**271-272**:155-161

[39] Muthukumar V, Suresh Babu A, Venkatasamy R, Senthil KN. An accelerated particle swarm optimization algorithm on parametric optimization of WEDM of die-steel. Journal of The Institution of Engineers (India): Series C. 2015;**96**:49-56

[40] Zhang G, Zhang Z, Ming W, Guo J, Yu H, Xinyu S. The multi-objective optimization of medium-speed WEDM process parameters for machining SKD11 steel by the hybrid method of RSM and NSGA-II. The International Journal of Advanced Manufacturing Technology. 2014;**70**:2097-2109

[41] Majumder A, Das A, Das PK. A standard deviation-based firefly algorithm for multi-objective optimization of WEDM process during machining of Indian RAFM steel. Neural Computing and Applications. 2018;**29**:665-677

[42] Giduturi S, Kuma A. Parameter optimization of wire EDM for H-13 tool steel. International Journal of Current Engineering and Technology. 2018;**8**:120-127

[43] Pamnani R, Vasudevan M, Vasantharaja P, Jayakumar T. Optimization of A-GTAW welding parameters for naval steel (DMR 249A) by design of experiments approach. Proceedings of the Institution of Mechanical Engineers, Part L: Journal of Materials: Design and Applications. 2017;**231**:320-331

[44] Choudhuri B, Sen R, Ghosh SK, Saha SC. Modelling and multi-response optimization of wire electric discharge machining parameters using response surface methodology and grey–fuzzy algorithm. Proceedings of the Institution of Mechanical Engineers, Part B: Journal of Engineering Manufacture. 2017;**231**:1760-1774

[45] Sudhakara D, Prasanthi G. Parametric optimization of wire electrical discharge machining of powder metallurgical cold worked tool steel using Taguchi method. Journal of the Institution of Engineers (India): Series C. 2017;**98**:119-129

[46] Reddy CB, Reddy CE, Reddy DR. Experimental investigation of surface finish and material removal rate of P20 die-tool steel in wire-EDM using multiple regression analysis. GSTF Journal of Engineering Technology (JET). 2017;**1**

[47] Manjaiah M, Rudolph FL, Kumar A, Basavarajappa S. Parametric optimization of MRR and surface roughness in wire electro discharge machining (WEDM) of D2 steel using Taguchi-based utility approach. International Journal of Mechanical and Materials Engineering. 2016;**11**:7. DOI: 10.1186/s40712-016-0060-4

[48] Tosun N, Cogun C, Inan A. The effect of cutting parameters on workpiece surface roughness in wire EDM. Machining Science and Technology. 2003;**7**:209-219

[49] Ugrasen G, Ravindra HV, Prakash GVN, Keshavamurthy R. Comparison of machining performances using multiple regression analysis and group method data handling technique in wire EDM of Stavax material. Procedia Materials Science. 2014;**5**:2215-2223

[50] Sudhakara D, Prasanthi G. Optimization of dimensional deviation: wire cut EDM of vanadis-4e (powder metallurgical coldworked tool steel) by Taguchi method. In: Proceedings of 5th International and 26th All India Manufacturing Technology, Design and Research Conference (AIMTDR-2014); 12-14 Dec 2014; IIT Guwahati: Assam. 2014

Chapter 3

Low Temperature Plasma Nitriding of Austenitic Stainless Steels

Tatsuhiko Aizawa

Additional information is available at the end of the chapter

http://dx.doi.org/10.5772/intechopen.78365

Abstract

A low temperature plasma nitriding process has become one of the most promising methods to make solid-solution hardening by the nitrogen super-saturation, being free from toxicity and energy consumption. High-density radio-frequency and direct current (RF/DC) plasma nitriding process was applied to synthesize the nitrided AISI304 microstructure and to describe the essential mechanism of inner nitriding in this low temperature nitriding (LTN) process. In case of the nitrided AISI304 at 673 K for 14.4 ks, the nitrided layer thickness became 66.5 μm with the surface hardness of 1550 HV and the surface nitrogen content of 9 mass%. This inner nitriding process was governed by the synergetic interrelation among the nitrogen super-saturation, the lattice expansion, the phase transformation, the plastic straining, the microstructure refinement, and the acceleration of nitrogen diffusion. When this interrelation is sustained during the nitriding process, the original austenitic microstructure is homogeneously nitrided to have fine-grained microstructure with the average size of 0.1 μm. Once this interrelation does not work anymore, the homogeneous microstructure changed itself to the heterogeneous one. The plastic straining took place in the selected coarse grains so that the parts of them were only refined. This plastic localization accompanied with the localized phase transformation.

Keywords: high-density plasma nitriding, low holding temperature (T_H), nitrogen super-saturation, γ-lattice expansion, phase transformation, elastic strain distortion, plastic straining, microstructure refinement, nitrogen diffusion path

1. Introduction

Iron-nitrogen or Fe-N binary system is essential in the steel design in a similar manner to the iron-carbon system [1]. Since the maximum solubility limit of nitrogen solutes is only 0.1

mass% or 0.3 at%, most of the previous studies concentrated on the solubility of nitrogen into γ-phase at a higher temperature than 1000 K [2]. Under those research circumstances, three important items in the material science were pointed out as findings to be noticed. Nitrogen works as a γ-phase stabilizer so that phase transformation temperature from α/α'- to γ-phase decreases with increasing the nitrogen content, [N] [3]. Mechanical properties significantly improve by themselves also with increasing [N] [4]. Thirdly, a crystalline structure of nitrogen super-saturated iron or Fe (N) resembles with that of iron nitride; for example, α'-Fe (N) with the nitrogen content of 11 at% has the same crystalline structure of α''-$Fe_{16}N_2$ and γ-Fe (N) with 20 at% nitrogen solute content corresponds to γ'-Fe_4N as surveyed in the textbook [5]. The first two items stimulated further researches for high-nitrogen steels (HNS) as discussed in [6]. The third item leads to the chemical vapor deposition of $Fe_{16}N_2$ thin films on the template substrate [7].

An pressurized electro-slag remelting (ESR) method has become a standard approach to fabricate an ingot of HNS [8]. Relatively high amount of dissociated nitrogen atoms from N_2 gas can diffuse into the depth of γ-phase matrix at 1263 K. This solution nitriding method induced the nitrogen atoms even into high chromium content steels including austenitic stainless steels [9]. Through this process, the nickel resource can be saved by 69 kg only by the addition of nitrogen atoms by 1 kg. The nitrogen works to stabilize the austenitic phase even with less nickel content. These nickel-free HNS have coarse grains, resulting in embrittlement, difficulty in welding, and insufficient stability in working. Although many trials have been made to improve the nitrogen solute content higher than 1 mass%, most of the studies experienced engineering difficulties to find a new alloying effect on the increase of strength and corrosion resistance [10].

In parallel with research on HNS, ion- and radical nitriding processes were developed with the use of the direct current (DC)-plasma and DC-pulse plasma technologies [11]. The T_H was held to be higher than 900 K in them; CrN was synthesized as a precipitate in the matrix together with iron nitrides such as Fe_2N and γ'-Fe_4N [12]. Hence, the stainless steels and Fe-Cr alloys were hardened by fine precipitation of CrN; however, the chromium content in the matrix was reduced by CrN-precipitation reaction to lower the original corrosion resistance [13]. In addition, this high temperature plasma nitriding was mainly governed by the nitrogen diffusion process; the nitrogen solute content exponentially decreases from the maximum nitrogen solid solubility of 0.1 mass% at the surface down to 0 toward the nitriding front-end [14]. Furthermore, when the T_H was higher than 1000 K, the chromium also diffuses to form a multi-stripe pattern with layered structure of {- (Cr-rich) - (CrN-rich) -} during nitriding [15]. Most of engineers and companies related to plasma nitriding believe that chemical reaction of chromium with nitrogen should drive the nitrided layer formation and hardening.

British research group [16] first found the nitrogen super-saturated lattices in the austenitic stainless steels by low temperature plasma nitriding. When using the same DC or DC-pulse plasmas, the nitriding process is characterized mainly by CrN precipitation into matrix when T_H is higher than 800 K. On the other hand, an original γ-lattice expands to form a peak shift from the original peaks of austenitic stainless steels in X-ray diffraction (XRD) analysis when $T_H < 800$ K. This finding does not mean a formation of new phase, so called by S-phase, but implies that nitrogen super-saturation accompanies with the γ-lattice expansion and that the crystalline structure of this Fe (N) is essentially different from that of original austenitic stainless steel matrix. In addition, various new engineering is expected to start from this nitrogen super-saturated Fe (N) [17].

In the present chapter, this low temperature nitriding (LTN) with nitrogen super-saturation is reconsidered by developing a new tool to drive LTN in the AISI304 stainless steels. First, Radio-frequency(RF)-DC plasma nitriding system is introduced with comments on the essential difference from other plasma nitriding processes such as DC- and DC-pulse plasmas. Quantitative plasma diagnosis equipment is stated to describe the nitrogen-hydrogen plasmas. In particular, the effect of hydrogen content in the mixture gas on the nitriding process is analyzed to determine the optimum condition. A hollow cathode device is proposed to intensify the ion and electron densities.

An austenitic stainless steel type AISI304 specimen is employed for plasma nitriding at 673 K for 14.4 ks by 60 Pa. Each fundamental process in this low temperature inner nitriding is analyzed by XRD, scanning electron microscopy (SEM)-electron dispersive X-ray spectroscopy (EDX), and electron back-scattering diffraction (EBSD). The γ-lattice expansion is analyzed as a peak shift in the XRD diagram. The nitrogen super-saturation is described by SEM-EDX; the elastic distortion is directly calculated by the lattice strain. The phase transformation, the plastic straining as well as the microstructure refinement are analyzed by EBSD. The nitrogen diffusion path is mainly estimated by the grain boundary diffusion process. These processes are mutually related to form a synergetic loop to drive this low temperature inner nitriding. When this loop is sustained during nitriding, the nitriding front-end advances homogeneously into the depth of stainless steel matrix. Once this loop is shut down at any point, the inner nitriding localizes by itself only to form a heterogeneous microstructure.

2. High-density plasma nitriding system

High-density RF-DC plasma nitriding system is introduced together with comments on the quantitative plasma diagnosis of nitrogen-hydrogen plasmas and on the hollow cathode device to intensify the ion and electron densities.

2.1. Background

DC-plasma and DC-pulse plasma [18–20] have been utilized for nitriding of stainless steel parts, tools and dies at higher hold temperature than 800 K. plasma enhanced chemical vapor deposition (PECVD) has been utilized for nitriding at lower temperature than 800 K. **Table 1** compares the difference in their capacity, inner nitriding behavior, and characteristics together with the present high-density RF-DC plasma nitriding. The former two approaches, widely utilized in the market, harden the stainless steels by CrN-precipitation and form thicker nitrided layer than 100 μm after 36 ks or 10 hours. PECVD nitriding works in the low-pressure of 1 to 2 Pa to form a moderate nitrided layer with the thickness of 20–30 μm. Inner nitriding process both in PECVD [21] and RF-DC plasma [22] is governed by nitrogen super-saturation without CrN precipitation. In the following, a detail of RF-DC plasma generation as well as a hollow cathode device is stated together with plasma diagnosis equipment in the present system.

Item	DC-Plasma (Ion Nitriding)	DC-Pulse Plasma (Radical Nitriding)	PECVD	High Density Plasma
Capacity	Large current High pressure	Small current High pressure	Small current Low pressure	Small current Marginable pressure
Holding Temperature (for SKD61)	823 K	823 K	673 – 693K	623 – 673 K
Inner Nitriding	CrN-precipitation	CrN-precipitation	Nitrogen Supersaturation	Nitrogen Supersaturation
Nitrided Layer Thickness	> 100μm for 36 ks	> 100μm for 36 ks	20—30μm for 10.8 ks	> 100μm for 14.4 ks

Table 1. Comparison of DC-plasma, DC-pulse plasma, PECVD, and RF-DC plasma nitriding processes.

2.2. RF-DC plasma nitriding system

A high-density plasma nitriding system [23–25] consisted of the vacuum chamber, the evacuation system, the DC-RF generators working in the frequency of 2 MHz, the gas supply of N_2 and H_2, and the heating unit located under the cathode plate as depicted in **Figure 1**. The nitriding parameters as well as controlling procedure are specified on the panels. All through the nitriding process, the measured pressure, temperature as well as gas pressure is automatically controlled by the process computer. Through the telecommunication, time history of RF- and DC-voltages and currents are also monitored on the panel to recognize the temporal status of RF-DC plasmas.

Figure 2 illustrates an actual experimental setup for plasma nitriding. The thermocouple is inserted into this cathode plate to monitor the T_H. In the vacuum chamber, the specimen is placed inside a hollow cathode setup on the cathode plate, which is electrically connected with DC generator. This hollow cathode setup includes a rectangle-shaped tube with the size of 40 × 20 × 70 mm^3 and the thickness of 2 mm. To be explained later, the specimen is located at the position with the highest nitrogen ion density, far half from the mixture gas inlet, as shown in **Figure 2**.

Figure 1. RF-DC plasma nitriding system. (1) Vacuum chamber, (2) RF-generator, (3) control-panel, (4) RF- and DC-power suppliers, (5) evacuation units, and (6) carrier gas supply.

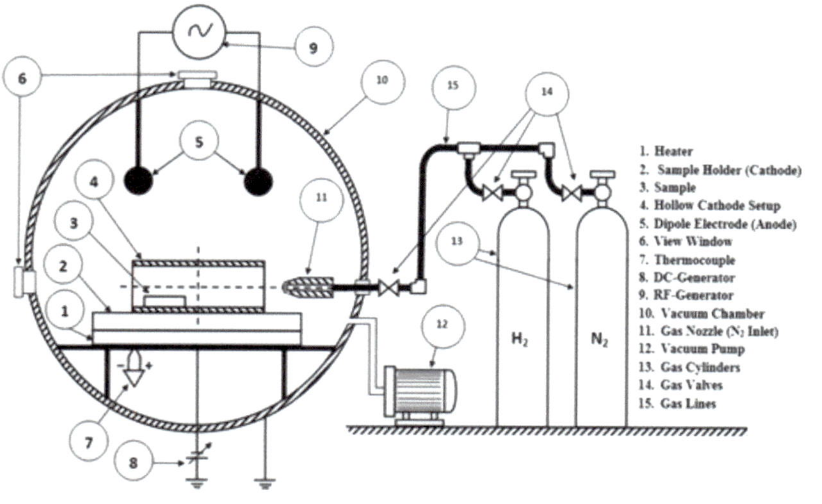

Figure 2. Experimental setup for RF-DC high-density plasma nitriding.

2.3. Quantitative plasma diagnosis

In the standard plasma diagnosis, two methods are often employed to quantitatively describe the nitrogen-hydrogen mixture plasma state; that is, emissive light optical spectroscopy (EOS) and Langmuir probe (LP). A typical setup for EOS is illustrated in **Figure 3** as precisely stated in the previous studies [26–28].

Emissive light from plasmas is detected through the optically transparent silica window on the chamber in **Figure 1** and analyzed to deduce the spectrum of species in the plasmas. **Figure 4** shows the effect of hydrogen content in the gas mixture on the measured spectra together with the spectra for hydrogen plasmas.

As seen in the spectra for hydrogen plasmas inserted in **Figure 4**, hydrogen peak intensity (Hα) at 656 nm increases monotonically with hydrogen content [H]. Pure nitrogen plasma mainly

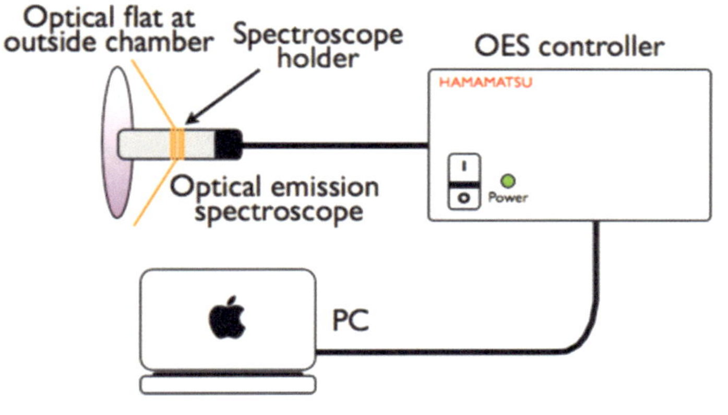

Figure 3. EOS for plasma diagnosis to describe the active species in the ignited plasmas.

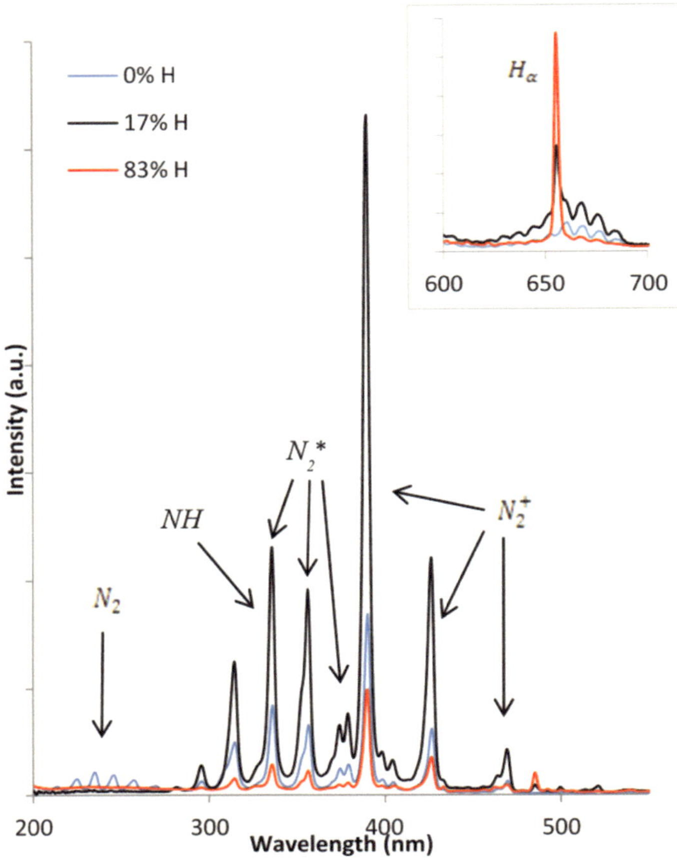

Figure 4. Measured spectra for the hydrogen-nitrogen plasmas by EOS.

consists of an excited nitrogen molecule (N_2*) and a nitrogen molecule ion (N_2^+) besides for a nitrogen molecule. When [H] = 83%, the whole population of nitrogen species including the NH-radicals significantly decreases. On the other hand, when [H] = 17%, {N_2*, N_2^+, and NH} have high-intensity without peaks for the molecular nitrogen in the range of short wavelength (<300 nm). Although the active species such as $N_2(A^3 \Sigma_u^+, \vartheta)$ or $N_2(X^1 \Sigma_g^+, \vartheta)$ cannot be observed directly from the EOS spectra, their transitions can be investigated to describe the reaction model on the active species; for example, $N_2(C^3 \Pi_u \rightarrow B^3 \Pi_{g*})$ peaks observed at 337, 358, and 370 nm are related to the N_2* and $N_2{}^+(B^2 \Sigma_u^+ \rightarrow X^2 \Sigma_g^+)$ at 391 and 427 nm, related to the N_2^+. These second positive and the first negative bands of nitrogen play an important role in the generation of atomic nitrogen by reaction with N_2 and N_2^+ in parallel with the formation of NH-radicals as detected at 336 nm.

The LP was also utilized in the diagnosis to describe the effect of [H] on the generated plasmas. **Figure 5** depicts how to measure the ion and electron densities by using LP.

Through the direct measurement of I-V curves at the probe tip, the electron resistivity as well as the ion and electron densities are analyzed to describe the plasma state. In particular, the electron resistivity is proportional to the enhancement of plasma chemical reaction. **Figure 6**

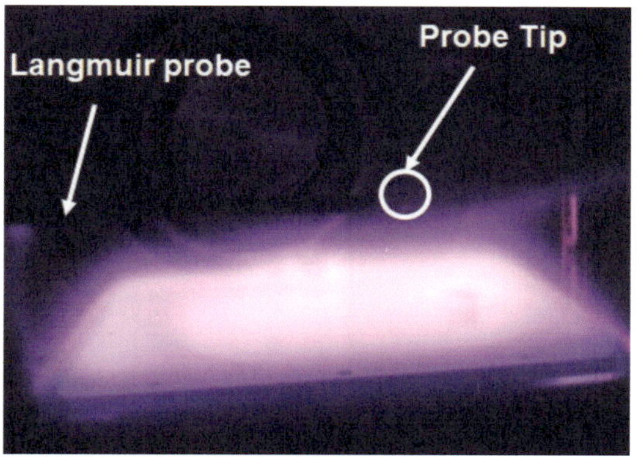

Figure 5. Experimental setups of the LP for measurement of ion and electron densities.

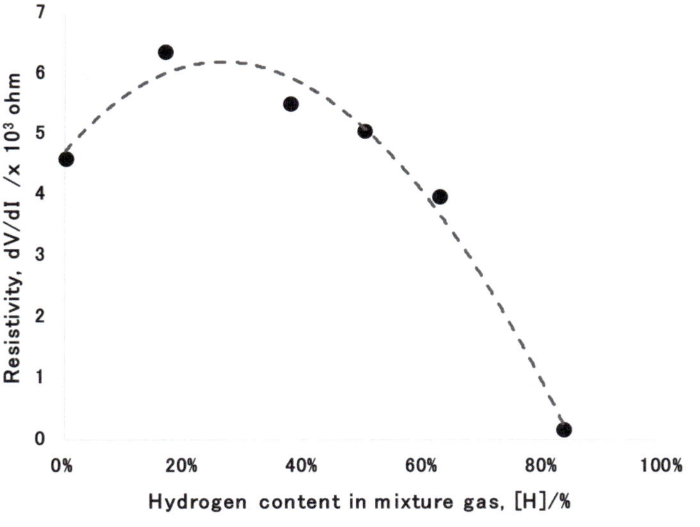

Figure 6. Variation of the measured resistivity in the plasmas with increasing the hydrogen content in the mixture gas.

shows a variation of measured resistivity in the plasmas with increasing the hydrogen content in the mixture gas.

The measured resistivity has maximum around [H] = 20–30%; a hot spot is formed in the plasmas where the chemical reaction is most enhanced for nitriding by directly controlling the hydrogen content [29].

2.4. Hollow cathode design for low temperature plasma nitriding

The hollow cathode device is utilized to intensify the ion density in the nitrogen-hydrogen plasmas. The LP is employed to directly measure the ion density in the hollow. The LP-tip was inserted into the hollow along the X-axis in every 2 cm. In each position, the tip was fixed at the center of hollow. **Figure 7** depicts the measured ion density distribution along the X-axis.

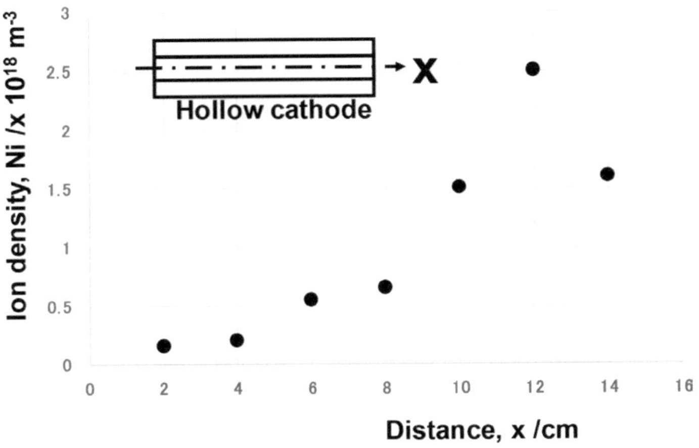

Figure 7. Measured ion density by the LP along the X-axis in the hollow.

The ion density increases monotonically with X in **Figure 7**; in particular, a hot spot with higher ion density than 1.5×10^{18} m^{-3} is located in the latter half of hollow. This is common to the hollow device effect where the ionization is enhanced at the vicinity of outlet in the hollow [30].

2.5. Observation and measurement

XRD (Rigaku SmartLab) with monochromatic Cu-Kα radiation (λ = 0.1542 nm) and Bragg–Brentano geometry, 40 kV, and 30 mA was utilized for analysis. The 2θ range was set between 30 and 90° with the scanning speed of 10 mm/min and the step angle of 0.02°. EDX device and software were utilized to make element mapping over a specified depth for nitrogen, chromium, iron, and carbon. Its spatial resolution was at most 5 μm. EBSD was utilized with the accelerating voltage of 20 kV, the working distance of 20 mm, the magnification (×2000), and the resolution of 0.1 μm. The inverse pole figure (IPF) was determined for each constituent grain to describe the change in microstructure through the nitriding. In addition, the kernel average misorientation (KAM) and the phase mapping were also measured to explain the plastic straining and phase transformation processes, respectively.

The micro-hardness testing apparatus (Mitsutoyo HM-200) was used by applying the load of 50 g or 0.5 N for hardness measurement on the cross-section in every 10 μm. The matrix hardness of AISI304 was 400 HV.

3. Low temperature plasma nitriding of austenitic stainless steels

An austenitic stainless steel type AISI304 was employed as a specimen for high-density plasma nitriding at 673 K for 14.4 ks by 60 Pa with use of the hollow cathode device. Essential processes in this low temperature plasma nitriding are described by chemical analyses.

3.1. Background

An austenitic stainless steel type AISI304 was selected for plasma nitriding at lower T_H than 723 K. **Table 2** summarizes the experimental results in the literature.

Although detail information is not written in a few papers, relatively high-nitrogen surface content and formation of a nitrided layer with the thickness of 10–20 μm are common to those previous studies [31–35]. More precise analysis and discussion are needed to investigate the essential processes, governing the inner nitriding behavior at a lower temperature than 700 K. High-density RF-DC plasma systems [36–40] provides a new way to further analyze this low temperature plasma nitriding by experiments.

3.2. Nitrogen super-saturation

In the present study, AISI304 stainless steel plate with the size of 40 × 20 × 2 mm^3 was employed as a specimen for RF-DC high-density plasma nitriding at 673 K for 14.4 ks by 60 Pa for the nitrogen and hydrogen mixture gas with the flow rate ratio of 160–30 ml/min. The pre-sputtering only with the use of nitrogen gas was first performed for 1.8 ks to clean the surface condition of the specimen before nitriding. **Table 3** lists the whole plasma nitriding condition in this experiment.

The nitrided specimen was halved to prepare the test-pieces for SEM-EDX analyses. **Figure 8** depicts the cross-sectional SEM image as well as the nitrogen mapping from the surface to the depth of nitrided specimen. The nitrogen content is measured at the surface to be 9 mass% or 26 at%. The nitrided layer thickness reaches to be 66.5 μm.

Compared with **Table 2**, both the nitrogen solute content at the surface and the nitrided layer thickness become the highest by using this plasma nitriding at 673 K for 14.4 ks. Formation of

References	Treatment systems	Stainless steels	Temperature (K)	Time (h)	Nitrogen Surface Concentration (at%)	Nitrided Thickness (μm)
E. Menthe, 1999	DC Plasma Nitriding	AISI304L	723 K	5 h	22 at%	11 μm
Y. Sun, 1999	DC Plasma Nitriding	AISI304	673 K	5 h	14 at%	10 μm
W. Liang, 2000	Plasma Nitriding	AISI304	693 K	1.5 h	17 at%	10 μm
L. Wang, 2006	Plasma Nitriding	AISI304	693 K	44 h	-	27 μm
S. Lu, 2017	DC Plasma Nitriding	AISI304	673 K	8 h	-	51.7 μm

Table 2. Previous studies on the low temperature plasma nitriding of AISI304 stainless steels.

Parameter	Pre-Sputtering	Plasma nitriding
Temperature (K)	673	673
Pressure (Pa)	300	60
RF-Voltage (V)	250	250
DC-Voltage (V)	350	500
Gas Flow (ml/min)	160 N_2	160 N_2, 30 H_2
Gas ratio (%)	100 %N_2	84%N_2+16%H_2
Duration (sec)	1800	14400

Table 3. RF-DC high-density plasma nitriding conditions.

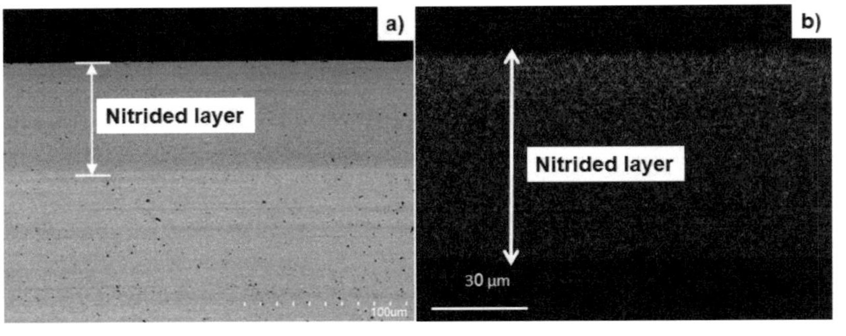

Figure 8. Cross-section of the plasma nitrided at 673 K for 14.4 ks. (a) SEM cross-sectional image and (b) nitrogen mapping from the surface to the depth.

uniform nitrogen super-saturated layer reveals that inner nitriding advances homogeneously into the depth of matrix.

3.3. Lattice expansion and elastic distortion

A nitrogen super-saturated lattice is expected to expand by itself; for example, the previous studies in **Table 2** reported a γ-lattice expansion by this nitrogen super-saturation. **Figure 9** compares the analyzed XRD diagrams before and after the plasma nitriding. The original austenitic phase is characterized by three peaks for γ (111), γ (200), and γ (220) detected at 2θ = 43.4, 50.82, and 74.5°, respectively.

Through the plasma nitriding process, all the austenitic lattices were elastically distorted to the expanded austenite (γ_N); for example, the original γ-peaks to AISI304 shifted to the lower 2θ, from 43.4 to 41.1°, from 50.82 to 47.94°, and from 74.5 to 70.08°, respectively. In parallel with these γ_N phases, the expanded martensitic peaks are also detected at 2θ = 43.7–63.5°. Negligibly small peak was detected at 2θ = 37.5° in trace level, which corresponds to the chromium nitrides. The γ-lattice expansion by this shift in XRD induces the tensile lattice strain by 5.4% for the peak shift from γ (111) to γ_N (111). This strain slightly increases to be

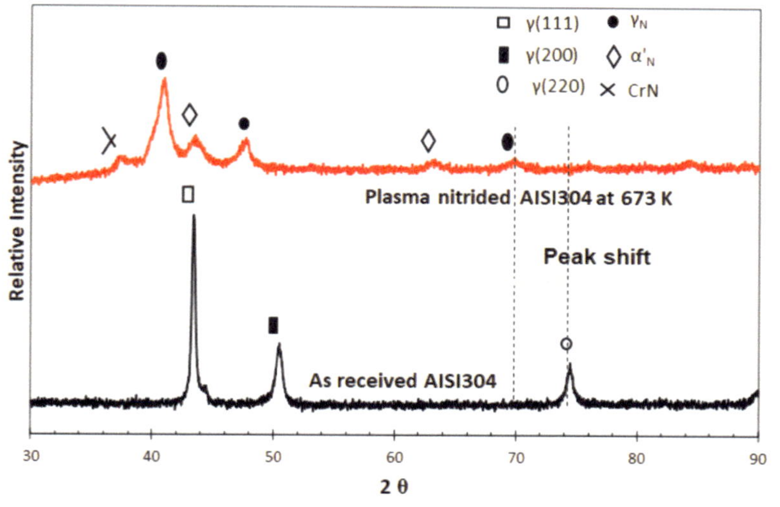

Figure 9. Comparison of analyzed XRD diagrams before and after plasma nitriding at 673 K.

5.6% for the shift of γ (200) to $γ_N$ (200) and 5.4% for γ (220) to $γ_N$ (220), respectively. This elastic distortion in the nitrogen super-saturated lattices just corresponds to the previous report in [41]. The grains housing these elastically distorted lattices are forced to deform plastically to compensate for strain incompatibility between the nitrogen unsaturated and the super-saturated lattices in grains.

3.4. Nitriding and hardening

EDX as well as the micro-Vickers testing are utilized to investigate the nitrogen content and hardness depth profiles. As depicted in **Figure 10**, the hardness becomes maximum at the

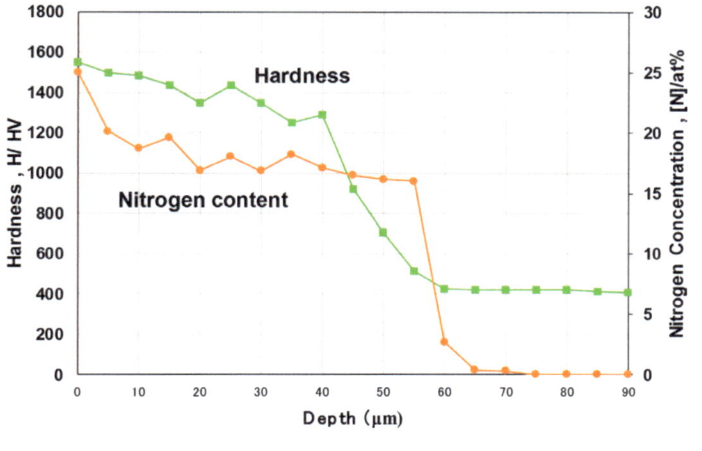

Figure 10. Nitrogen content and hardness depth profiles from the surface to the depth.

surface by 1550 HV and gradually reduces down to 1300 HV in the depth of d < 40 μm. From d = 40–66.5 μm, this hardness gradually decreases to the matrix hardness of 400 HV. A nitriding front-end is defined by the position in depth where the measured hardness coincides with the substrate hardness; the nitrided layer thickness (E) after nitriding for 14.4 ks is 66.5 μm. The nitrogen content is kept constant to be 15–17 at% in the depth of 5 < d < 60 μm except for the vicinities of surface and nitriding front-end. In the high temperature nitriding, the nitrogen content exponentially decays from the maximum nitrogen solubility limit of 0.1 mass% at the surface and goes to zero at the nitriding front-end. The nitrogen solute content depth profile in **Figure 10** is far from the common knowledge on the inner nitriding process in the high temperature nitriding.

3.5. Phase transformation and plastic straining

EBSD provides a tool to describe the interrelation among the phase transformation, the plastic straining, and the microstructure evolution. The measured phase mapping and KAM distribution on the cross-section of the nitrided specimen are shown in **Figure 11a** and **b**, respectively. After [42], the measured cross-sectional KAM profile can be identified as an equivalent plastic strain distribution.

In **Figure 11a**, the expanded γ-phase and transformed α'-phase finely distribute in the depth of d < 40 μm and form the two-phase microstructure. This homogeneous two-phase structure abruptly changes to a heterogeneous one where α'-phase sparsely distributes in the γ-phase matrix. This autonomous phase mapping change coincides with the onset of hardness reduction in **Figure 10**. This is because the volume fraction of extended γ-phase and transformed α'-phase zones begins to reduce from d = 40 μm in **Figure 11a**.

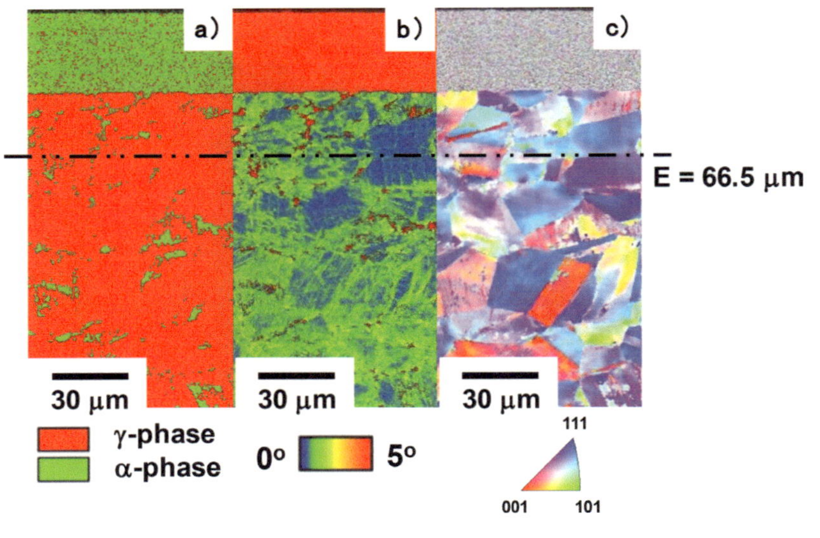

Figure 11. EBSD analysis on the cross-section of the nitrided AISI304 specimen at 673 K for 14.4 ks. (a) KAM distribution, (b) phase mapping, and (c) IPF depth profile.

In **Figure 11b**, the layer of d < 40 μm is plastically strained in all since most of grains house the expanded lattice zones and plastically distort to compensate for the strain incompatibility in each grain. Just as seen in **Figure 11a**, this homogeneous plastic straining also changes by itself and localizes to specified grains. In fact, plastic strains localize in each grain below d = 40 μm. That is, uniform phase transformation and plastic straining change themselves across this critical depth by their localization to grains. The neighboring lattices to elastically distorted ones by phase transformation are forced to make plastic distortion. The transformed α'-phase zones in **Figure 11a** correspond to the highly strained zones in **Figure 11b**.

3.6. Microstructure refinement

Phase transformation and plastic straining in the above reflects the microstructure change by the nitrogen super-saturation. **Figure 11c** depicts the IPF distribution on the cross-section of the nitrided specimen. Each grain with a specified crystallographic orientation is represented by a different color. In correspondence to **Figure 11a** and **b**, the layer of d < 40 mm has refined microstructure with the average grains size of 0.1 μm, just near the spatial resolution of EBSD. This gray color for this layer in **Figure 11c** implies that each grain in this layer is homogeneously refined to have random crystallographic orientation. Just as observed in **Figure 11a** and **b**, this homogeneous microstructure changes by itself to heterogeneous one at d = 40 μm; the average grain size comes near to the original grain size before nitriding. To be noticed, the crystallographic orientations with different colors from original one or with graded colors are seen in most of the grains below d = 40 μm. The plastically strained grains are partially decomposed into several or tens of subgrains with different crystallographic orientations.

3.7. Summary

LTN of austenitic stainless steels is essentially different from the conventional plasma nitriding at higher temperatures. No nitrides are formed in the matrix so that no change in the original chromium content proves less change in the original corrosion resistance of stainless steels. The surface layer is hardened by nitrogen solid-solution where the γ-phase is expanded by nitrogen super-saturation with the occupation of octahedral vacancy sites by nitrogen solutes. Owing to fine grain size in the homogeneously nitrided layer, higher strength is expected to this high-nitrogen stainless steel surface. In addition, the fine-grained two-phase structure has a role to improve the trade-off-balancing between strength and fracture toughness and to increase the fatigue life [2]. How to extend this homogeneously nitrided layer toward the nitriding front-end must be an engineering issue to be discussed further.

4. Discussion

Inner nitriding mechanism in this low temperature plasma nitriding of austenitic stainless steels is discussed with importance on the difference between the homogeneous and heterogeneous nitriding processes.

4.1. Synergetic loop to drive the low temperature plasma nitriding

LTN mechanism is described by a synergetic loop as explained by **Figure 12**. Nitrogen solute, penetrating from the surface under high-nitrogen flux, occupies with an octahedral vacancy sites in the fcc-structured lattice as suggested by [43]. Under this nitrogen super-saturation, the γ-lattice expands, and elastically distorts to drive the γ to α' phase transformation. The whole γ-lattices neighboring to expanding γ-lattices and transformed α'-lattices, are plastically strained to compensate for the strain incompatibility between two zones. Original grain is distorted and decomposed into fine subgrains by this plastic straining. More nitrogen solutes diffuse to the depth of unsaturated matrix through the refined grain boundaries. Evolution of the nitrided layer accompanies with this loop.

4.2. Homogeneous nitriding process

When this synergetic loop is sustained during the plasma nitriding, every unit process uniformly advances from the surface to the depth of matrix. As seen in **Figure 11**, the original coarse-grained AISI304 matrix is surface-modified to have fine-grained, two-phase microstructure from the surface to the depth of 40 μm. This homogeneous nitriding is shut down at the critical depth of 40 μm for the nitrided AISI304 at 673 K for 14.4 ks in **Figure 11**; the above loop only takes place locally below this critical depth. There is no change in the synergetic loop across this criticality. When the loop works uniformly in the matrix, the nitriding advances homogeneously, while it does heterogeneously when the loop localizes in the selected grains. In other words, this autonomous change from heterogeneous nitriding to homogeneous nitriding is driven by the nitrogen super-saturation process into grains. When the nitrogen super-saturated γ-phase zones are closely neighboring to each other in the specified grain, the whole related grains are homogeneously nitrided and refined by the synergetic loop in **Figure 12**. On the other hand, when each super-saturated γ-phase zones are isolated from each other, every process in the loop works only inside of each grain.

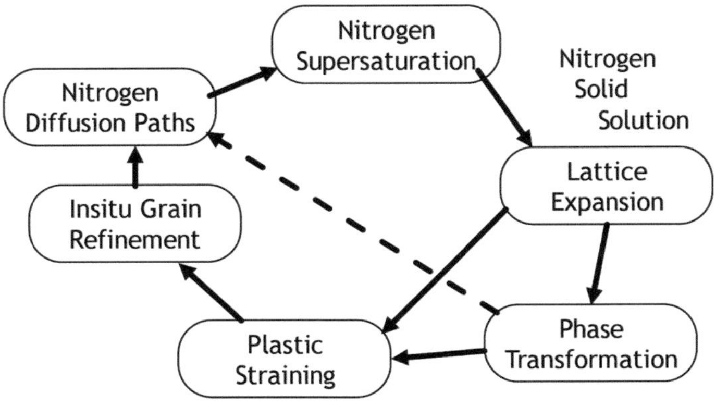

Figure 12. Synergetic loop of processing steps to drive the low temperature inner nitriding of stainless steels.

4.3. Heterogeneous nitriding process

Let us first describe the localization of phase transformation, plastic straining, and micro-refinement below the critical depth in **Figure 11**. The phase mapping, the plastic strain distribution as well as the IPF mapping for the grain-A at d = 100 μm in **Figure 11** are analyzed and shown in **Figure 13**. Since a grain boundary works as a nitrogen diffusion path, most of the γ-phase zones at its vicinity transform to α'-phase. As pointed by an arrow-a in **Figure 13a** and **b**, a series of α'-phase zones are aligned in the alternate order of {- (α'-phase zone) – (highly plastic-strained zone) – (α'-phase) -}. In correspondence to this alignment, an original (001) orientation rotates by the plastic straining as shown in the graded colors in **Figure 11c**. The transformed zones have (111) orientation as pointed by arrow-b. This local change in phase mapping, plastic straining, and crystallographic orientation distribution in the inside of grain-A proves that the heterogeneous nitriding process is driven by this localization in each grain of matrix below the critical depth.

Figure 13 also suggests that each transformed band, pointed by the arrow-a, has a unit size of 0.3–0.5 μm in common and that these bands are isolated by highly strained γ-phase zones. With the enhancement of the nitrogen flux from the surface or with an increase of the nitrogen diffusion path density, those isolated zones overlap with each other to change the heterogeneous nitriding to the homogeneous nitriding. In other words, homogeneous nitriding mode prevails in the low temperature plasma nitriding process with higher activation of nitrogen flux from plasmas or with reduction of the initial grain size to a comparable level of transformed units in **Figure 13**.

4.4. Initial grain size effect on nitriding process

An initial grain size of AISI304 sheet is controlled to decrease by intense rolling with the reduction in thickness by 90 % to demonstrate this mode change from heterogeneous nitriding to

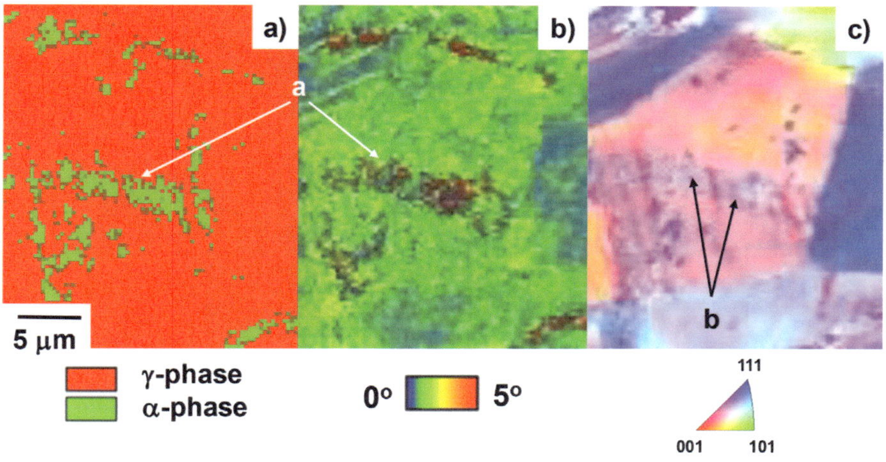

Figure 13. Localized steps around the grain-a in the heterogeneous nitriding process. (a) Local phase mapping, (b) local plastic straining, and (c) local IPF mapping.

homogeneous nitriding. **Figure 14** depicts the phase mapping, the plastic straining, and the microstructure refinement on the cross-section of rolled AISI304 before nitriding. Although crystallographic textures are formed along the rolling direction as shown by the arrow-a in **Figure 14**, the average grain size is uniformly reduced down to 1.5 μm.

Under the same processing conditions, this fine-grained AISI304 specimen is nitrided at 673 K for 14.4 ks. The nitriding front-end is analyzed to be E = 60 μm, nearly the same as shown in **Figure 11**. **Figure 15** shows the phase mapping, the plastic strain distribution, and the microstructure after nitriding. The heterogeneous microstructure observed above the nitriding front-end as well as the textures by rolling in **Figure 14**, completely disappears to form two-phase and fine-grained homogeneous nitrided layer. Although the initial fully martensitic phase turns to be $\gamma - \alpha'$ two-phase; this two-phase fine microstructure is continuously formed across the nitriding front-end. This homogenization of the heterogeneous microstructure reveals that nitrogen super-saturation process advances homogeneously into the depth of stainless steel matrix under the synergetic loop once the grain size of the matrix is comparable to the nitrogen super-saturated unit size.

4.5. Summary

LTN of stainless steels is essentially governed by the homogeneous nitrogen super-saturation. When the synergetic loop is sustained during the nitriding, the nitrided layer has two-phase, fine-grained microstructure. Once the nitrogen super-saturation process is localized into the specified grains, the homogeneous nitriding changes itself to heterogeneous nitriding.

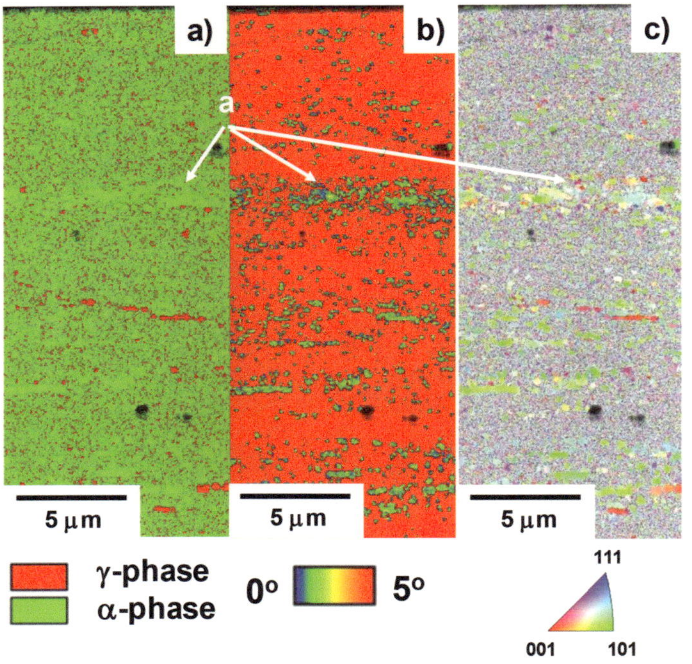

Figure 14. Microstructure of intensely rolled AISI304 sheet with the reduction of thickness by 90%. Average grain size is 1.5 μm.

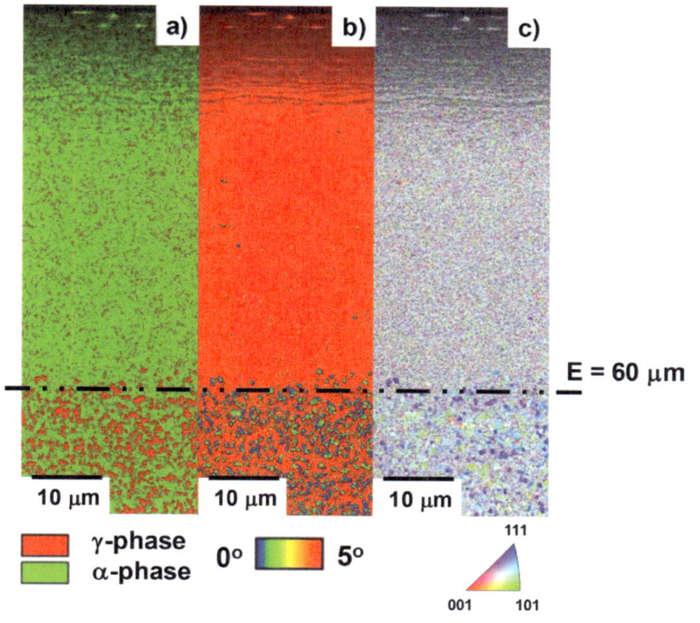

Figure 15. Microstructure after nitriding the fine-grained AISI304 at 673 K for 14.4 ks.

Refinement of the initial grain size into a comparable size of nitrogen saturated γ-lattice units homogenizes the heterogeneously nitrided layer.

5. Conclusion

Low temperature plasma nitriding provides a processing tool for the surface treatment of the stainless steels to have a fine two-phase microstructure with the average grain size of 0.1 μm. This homogeneously nitrided layer has higher surface hardness than 1500 HV and higher nitrogen content than 15 at%. Different from the conventional nitriding, (1) no nitrides are formed as a precipitate in the matrix, (2) stainless steel matrix is nitrogen super-saturated to have a nitrogen content plateau of 15–17 at% toward the nitriding front-end, (3) phase transformation and plastic straining take place together with γ-lattice expansion, and (4) original coarse grains are refined. This homogeneous nitriding follows the heterogeneous nitriding process where γ-lattice expands locally in the specified grains with phase transformation to α'-phase and plastic straining. When nitriding the fine-grained stainless steels, their surfaces are homogeneously nitrided to have fine, a two-phase microstructure with high hardness, strength, and corrosion resistance.

Acknowledgements

The author would like to express his gratitude to Mr. Abdelrahman Farghali (SIT) and late Mr. Y. Sugita (YS-Electric Industry, Co. Ltd.) for their help in experiments. This study was financially supported in part by the Abe-Initiative in Japan Government and the METI-program on the supporting industries, Japan, respectively.

Conflict of interest

No conflicts of interest were declared.

Author details

Tatsuhiko Aizawa[1,2*]

*Address all correspondence to: taizawa@sic.shibaura-it.ac.jp

1 Shibaura Institute of Technology, Tokyo, Japan

2 Surface Engineering Design Laboratory, LLC, Tokyo, Japan

References

[1] Massalski TB, editor. Binary Alloy Phase Diagrams, Volume 2. Handbook on Phase Digrams. 2nd ed. ASM International; 1990. p. 1728

[2] Imai Y, Izumiyama M, Tsuchiya M. Thermodynamic study on the transformation of austenite into martensite in iron-high nitrogen and iron carbon binary system. Journal of the Japanese Institute of Metals and Materials. 1965;**29**:427-435

[3] Bell T, Owen WS. Journal of the Iron and Steel Institute. 1967;**205**:428

[4] Gavriljuk V et al. In: Proceedings of 1st High Nitrogen Steels; Lilly; 1988. p. 447

[5] Murata I, Sakamoto M. Properties of High Nitrogen Steels. Agune; 2005. pp. 166-168

[6] Katada Y. Outlook 2006 of National Institute of Materials Sciences. Tsukuba: National Institute of Materials Sciences; Cap3-5-72006. pp. 342-345

[7] Komuro M, Kozono Y, Hanazono M, Sugita Y. Structure and saturation magnetic flux density of epitaxially grown Fe and Fe-N films on GaAs (100). 1989;**13**(2):301-306

[8] W Holzgruber. 1974. Austrian Patent No. 333327, S12

[9] Stein G, Menzel J, Dorr H. In: Foct J, Hendry A, editors. High Nitrogen Steels HNS88. London: Maney Publishing; 1988. p. 32

[10] Berns H, Kühl A. Reduction in wear of sea page pump through solution nitriding. Wear. 2004;**246**:16-20

[11] Aizawa T, Sugita Y. High density RF-DC plasma nitriding of steels for die and mold technologies. Research Report. SIT. 2013;**57**(1):1-10

[12] Aizawa T, Sugita Y. High density plasma nitriding of tool and die steels. In: Proceedings of the 6th South East Asian Technical University Consortium Symposium; 2012. pp. 134-137

[13] Kuwahara H. Surface modification of iron alloys by plasma nitriding and carburizing [PhD Thesis]. Kyoto University; 1992

[14] Hiraoka Y, Inoue K. Prediction of nitrogen distribution in steels after plasma nitriding. Denki-Seiko. 2010;**86**:15-24

[15] Granito N, Kuwahara H, Aizawa T. Normal and anormal microstructure of plasma nitrided Fe-Cr alloys. Journal of Materials Science. 2002;**37**(4):835-844

[16] Bell T. Surface engineering of austenitic stainless steel. Surface Engineering. 2002;**18**: 415-422

[17] Dong H. S-phase surface engineering of Fe-Cr, Co-Cr and Ni-Cr alloys. International Materials Reviews. 2011;**55**(2):65-98

[18] Saito K. Introduction to Ion Nitriding in Hokunetsu. 2018. p. 3. http://hokunetsu.com/products/003/

[19] Gruen P. PulsePlasma@. Broacher of PlaTeg; 2008

[20] Granto N. Surface material design and control in steels for automotive parts by plasma nitriding technique [PhD thesis]. University of Tokyo; 2003

[21] Marcos G, Guilet S, Cleymand F, Thiriet T, Czerwiec T. Stainless steel patterning by combination of micro-patterning and driven strain produced by plasma assisted nitriding. Surface and Coatings Technology. 2011;**205**:5275-5279

[22] Aizawa T, Fukuda T, Itoh K. Duplex coating of AISI316 and 420 dies for holt mold-stamping. Journal of Iron and Steel International. 2012;**219**:933-936

[23] Katoh T, Aizawa T, Yamaguchi T. Plasma assisted nitriding for micro-texturing onto martensitic stainless steels. Manufacturing Review. 2015;**2**:2, 1-7

[24] Aizawa T, Fukuda T, Itoh K. Duplex coating of AISI-316 and 420 dies for hot mold-stamping. Steel Research International Journal. 2012;**212**:933-936

[25] Istiroyah, Aizawa T, Santjojo DJ. High density plasma nitriding behavior of austenitic stainless steel type AISI 316. In: Proceedings of the 7th SEATUC Conference; Bandon; 2013 (CD-ROM)

[26] Djoko DJ, Aizawa T. Formation of expanded martensite in plasma nitrided AISI420 stainless steel. In: Proceedings of the 8th SEATUC Conference; Johor-Balu; 2014 (CD-ROM)

[27] Yunata EE. Characterization and application of hollow cathode oxygen plasma [PhD thesis]. SIT; 2016

[28] Rosadi I, Djoko S, Sakti SP, Yunata EE, Aizawa T. Plasma diagnosis on the mixture gas plasma states for nitriding and surface activation. In: Proceedings of the 12th SEATUC Conference; Yogyakarta, Indonesia; 2018 (in press)

[29] Santjojo D, Itsiroyah, Aizawa T. Dynamic of nitrogen and hydrogen species in a high rate plasma nitriding of martensitic stainless steels. In: Proceedings of the 9th SEATUC Conference; Thai; 2015 (CD-ROM)

[30] Yunata EE, Aizawa T, Santojojo DJ. Characterization of hollow cathode plasma for ashing processes. In: Proceedings of the 8th SEATUC Conference; Johor-Balu; 2014 (CD-ROM)

[31] Menthe E, Rie K-T. Further investigation of the structure and properties of austenitic stainless steel after plasma nitriding. Surface and Coatings Technology. 1999;**116**:199-204

[32] Sun Y, Li XY, Bell T. X-ray diffraction characterization of low temperature plasma nitrided austenitic stainless steels. Journal of Materials Science. 1999;**34**:4793-4802

[33] Liang W, Xiaolei X, Bin X, Zhiwei Y, Zukun H. Low temperature nitriding and carburizing of AISI304 stainless steel by a low pressure plasma arc source. Surface and Coatings Technology. 2000;**131**:563-657

[34] Wang L, Ji S, Sun J. Effect of nitriding time on the nitrided layer of AISI 304 austenitic stainless steel. Surface and Coatings Technology. 2006;**200**:5067-5070

[35] Lu S, Zhao X, Wang S, Li J, Wei W, Hu J. Performance enhancement by plasma nitriding at low gas pressure for 304 austenitic stainless steel. Vacuum. 2017;**145**:334-339

[36] Aizawa T. Functionalization of stainless steels via low temperature plasma nitriding. In: Proceedings of the 7th Annual Basic Science International Conference; Malang, Indonesia; 2017. pp. 1-16

[37] Farghali A, Aizawa T. Phase transformation induced by high nitrogen content solid solution in the martensitic stainless steels. Materials Transactions. 2017;**58**:697-700

[38] Farghali A, Aizawa T. Nitrogen supersaturation process in the AISI420 martensitic stainless steels by low temperature plasma nitriding. ISIJ International. 2018;**58**(3):401-407

[39] A. Farghali, T. Aizawa, Homogeneous and Heterogeneous Structuring Processes in the Nitrided Austenitic Stainless Steels below 673K. Materials Science and Engineering, (2018) (in press)

[40] Aizawa T, Yoshihara S. Homogeneous and heterogeneous micro-structuring of austenitic stainless steels by the low temperature plasma nitriding. In: Proceedings of 7th International Conference on Material Strength and Applied Mechanics; Kita-Kyushu, Japan; 2018 (in press)

[41] Öztürk O, Williamson DL. Phase and composition depth distribution analyses of low energy, high flux N implanted stainless steel. Journal of Applied Physics. 1995;**77**: 3839-3850

[42] Kamaya M, Wilkinson AJ, Titchmarsh JM. Quantification of plastic strain of stainless steel and nickel alloy by electron backscatter diffraction. Acta Materialia. 2006;**54**:539-548

[43] Domain C, Becquart CS, Foct J. Ab initio study of foreign interstitial atom (C, N) interactions with intrinsic point defects in α-Fe. Physical Review B. 2004;**69**:144122

Chapter 4

HAZ Phase Transformation and Thermal Damage for Laser Remanufacturing a High-Strength Stainless Steel

Shi-yun Dong, Xiang-yi Feng, Jin-xiang Fang and
Shi-xing Yan

Additional information is available at the end of the chapter

http://dx.doi.org/10.5772/intechopen.79910

Abstract

It briefly introduced laser remanufacturing, which was an advanced repairing method to refabricate damaged components based on laser forming technologies. The possible factors in determining the performance of the laser remanufacturing FV520B were studied by numerical simulation and experimental methods. First, the results of free dilatometry test showed that the volume effect of phase transformations were corresponding to the transformation temperatures and heating rate of the laser process had remarkable effects on the kinetics of phase transformation. In addition, the evolution of temperature fields of the single-pass and multi-layer laser cladding processes were analyzed by numerical simulation method based on deactivate and reactivate element theory. A combined method of dilatometry and metallography was conducted to reveal the effect of cooling condition and phase transformation on the microstructure of HAZ. The maximum temperature of thermal cycle had a dominating effect on the microstructure, microhardness and phase transformation temperature rather than cooling rate. Thermal cycles had a significant effect on the metallographic transformation and consequently decided the mechanical performance. Microhardness and tensile tests were conducted and the results showed that strength and ductility of laser remanufacturing FV520B were equivalent to that of forgings.

Keywords: phase transformation, laser remanufacturing, heat-affected zone, martensite stainless steel

1. Introduction

Conventional high strength steel (CHSS) has been defined in advanced high-strength steels (AHSS) Application Guideline 3.0 by World Steel Association (IISI) as steel material with high strength (yield strength ≥ 210 MPa and tensile strength ≥ 370 MPa). In recent years, light alloy like aluminum and magnesium alloys were widely used in automobile lightening and it also promotes the emergence and development of the AHSS. As a typical example of high strength steels, martensitic stainless steel with good comprehensive mechanical properties, casting performance and excellent corrosion resistance was used in various large-scale and complex operating conditions equipment, such as large-scale compressor rotors, air-compressor blade and structural components of the nuclear reactor. These high performance parts with complicated structure and high manufacturing cost had great potential to remanufacture. The suitable material testing technique and perfect remanufacturing process were desperately needed for the remanufacturing of damaged equipment and it is environmentally friendly, more economical and efficient.

The laser cladding processing is an advanced high-energy beam remanufacturing techniques. As the light source, laser beam had high energy-density, high-precision and high flexible processing to achieve net shaping forming for manufacturing those structural parts or components with complex inner structure [1]. Taking full advantages of laser beam and based on the laser cladding, laser remanufacturing technology was used for remanufacturing the disabled parts of the damaged equipment and the original shape or external dimensions recovered accurately after proper subsequent machining. Meanwhile, the alloy powder material and the matrix of the work-piece produce high quality metallurgically bonding under heating of powerful laser during the remanufacturing processing and the overall mechanical properties of the remanufacturing parts possibly can meet or exceed that of new one by the regulation of the microstructure and the mechanical properties of heat affected zone and cladding. In general, as an advanced refabrication technology, laser remanufacturing was widely applied in repairing and regeneration of many high value-added equipment [2–6].

Although laser remanufacturing technology had been widely applied in many industries like aerospace, oil drilling, electricity, automobile and machine, there exist some unsolved problems causing seriously restricted the development and use of this technology. Quite different with the laser additive manufacturing process, the deposition of the alloy powder material was on the surface of the work-piece during the laser remanufacturing and therefore the interface between the cladding and the matrix cannot be neglected. Two specific aspects of the work were completed during the laser remanufacturing. Firstly, the high-density and high mechanical properties cladding part was obtained by the design and control of the laser cladding alloy powder material, the process optimization and the proper postprocessing like stress-relief heat treatment, which is similar to the laser additive manufacturing process. However, the regions adjacent to the surface of the work-piece (the depth of the region was about 100–1000 micrometers according to many studies) experienced different short multi-cycle with superhigh heating and cooling rate and moreover a series of nonequilibrium solid state phase transformation occurred during the interaction between laser beam and alloy powder material. Consequently, the microstructure of the heat affected zone was different with that of the work-piece before remanufacturing and the mechanical performance may deteriorate, namely thermal damage during the laser remanufacturing. Recently, the 3D print technology had drawn a lot of attention of large enterprises and academic institutions. Studies on the laser additive manufacturing processing optimizing were done by the combination of numerical simulation and experimental methods. Some mechanism of the grain morphology control was investigated by basic study on solidification nucleation and growth mechanisms of the local melting pool. Most of the studies focused on the pretreatment and heat-treatment during the laser cladding process and it turned out that the microstructure and mechanical properties improved significantly. Consequently, heat affected zone (HAZ) was the weak region for the laser remanufacturing parts and the relevant issues had deep effects on the development of this technology [7–11].

Until now, studies on the mechanism of the microstructure evolution and mechanical properties distribution were sorely lacking. The key parts of the issues were the nonequilibrium phase transformation and the evaluations of the mechanical properties of HAZ. First of all, the regions in different depth in HAZ experienced different thermal cycles and eventually the microstructure and the microhardness of HAZ show a stepwise characteristic. Besides, the microstructure of HAZ may have further changes as the following and continuous heat cycles. For some material systems like martensite stainless steel, a set of characteristic parameters (maximum temperature, heating rate and cooling rate) of the heat cycles in HAZ can be acquired by methods like computer numerical simulation and temperature-measuring technology. To some extent, the formation mechanism of the final microstructure in HAZ can be explained and predicted by the classic phase transformation theory. However, some none-equilibrium microstructure frequently appear in HAZ, which make controlling the microstructure and the mechanical properties challenging. In addition, the mechanical properties evaluation of HAZ was challenging but significant. According to previous studies, by prefabricated delimitation the butt joint specimens were obtained as the mechanical properties tests samples or obtained directly from the laser cladding part along the cladding direction.

2. Solid state phase transformation during the remanufacturing processing of martensite stainless steel

Martensite phase transformation was a displacive phase transformation. During the martensite phase transformation, lattice type were changed by shear of material atoms, as well as the mechanical, thermodynamic properties and specific volume of the material. The thermodynamic and latent heat of phase transformation changes will have massive effects on the evolution of the temperature field.

During the laser cladding, material will experience a complete cycle from room temperature up to melting points. The physical properties of material changed as the occurrence of the solid state phase transformation and then eventually affected the evolution of the temperature field and stress field. The studies on the temperature and stress field evolution during the laser remanufacturing were based on the reveal of physical properties evolution. Furthermore, the heat cycles of different regions in HAZ can be obtained from the temperature field evolution and then the mechanism of the microstructure and mechanical properties distribution in HAZ may be revealed. Last but not the least, we hope the final microstructure and mechanical properties of HAZ can be accurately predicted and well controlled. However, the solid state phase transformation during the laser cladding was a non-equilibrium phase transformation and the kinetics of phase transformation were obviously affected by the superhigh heating rate and cooling rate [12, 13].

2.1. The effects of solid state phase transformation on the specific volume of martensite stainless steel

In the case that no phase transformation occurs for some generic solid material, the temperature rises will increase as the internal energy rises, so is the volume. On the contrary, the temperature and the volume decrease as the internal energy fall.

The specimens of the thermal expansion coefficient measurement were shown as **Figure 1**. The dilatometry experiment was processed in DIL801 dilatometer (**Figure 2**). In the temperature range (−150~1500°C), the maximum heating rate was 50°C/min and the maximum cooling rate was 25°C/min at low temperatures. In addition, the maximum heating rate rise to 100°C/min and the maximum cooling rate reached to 180°C/min as well. The errors of linear expansion coefficient was set as 0.03*10–6/°C, the gradient of temperature was 10°C/min. Using nitrogen as carrier gas, the flow rate was set as 20 ml/°C. The ambient temperature was 25°C and humidity was 35%.

The radial strain of the specimens for the thermal expansion coefficient experiments were used as measurement results. The strain was decomposed into two parts in the Ti (Range from room temperature to Ms): One part were the strain variations of martensite and austenite as temperature changes. The rest were caused by the volume effect of martensite phase transformation. It could be described as follows:

$$\Delta L/L_0 = \beta_M^0 V_i + (1 - V_i)\, \alpha_A T_i + V_i \alpha_M T_i$$

dealed as:

$$\beta_M^0 = (\alpha_\alpha - \alpha_\gamma)\Delta T - \Delta \varepsilon_M^{RT}$$

where $\Delta \varepsilon_M^{RT}$ are the intercept of the extension line of austenite section during the cooling period in the temperature-strain curves and β_M^0 are the strain caused by the volume effect of solid state phase transformation.

For FV520B high-strength stainless steel, the free expansion curve was shown as **Figure 3**. The critical temperature of transformation from pearlite to austenite (Ac1) was 600°C and the complete transformation temperature of austenite (Ac3) was 900°C. The starting temperature of the martensite phase transformation (Ms) was 130°C.

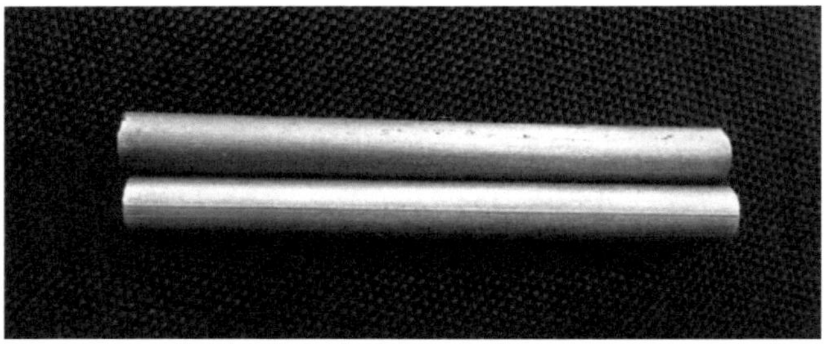

Figure 1. Specimens for the free dilatometry test.

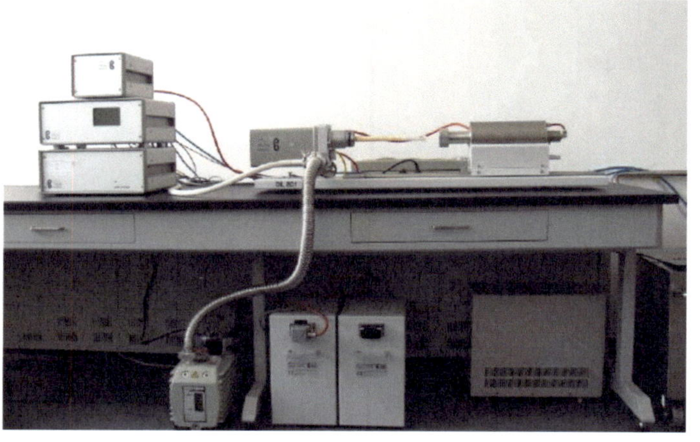

Figure 2. DIL801 single-sample dilatometer.

Shown as **Figure 4**, in the temperature between 130 and 600°C, the strain of martensite was 0.0057 and that of austenite was 0.088. The expansion coefficient of austenite was 18.72 and that of martensite was 12.13. In high temperature, the expansion coefficient of austenite was 21.5 and that of martensite reached up to nearly 19 at temperature over 600°C.

Figure 5 shows the temperature range of martensite transformation during the free dilatometry test. In the martensite phase transformation period, the strain caused by volume effect during phase transformation was 0.0067249 and the real strain should be 0.00328 considering the thermal contraction. Ideally, the material volume increased by 0.97% compared with that at Ms. In addition, the material volume at 320°C was equivalent to that at room temperature.

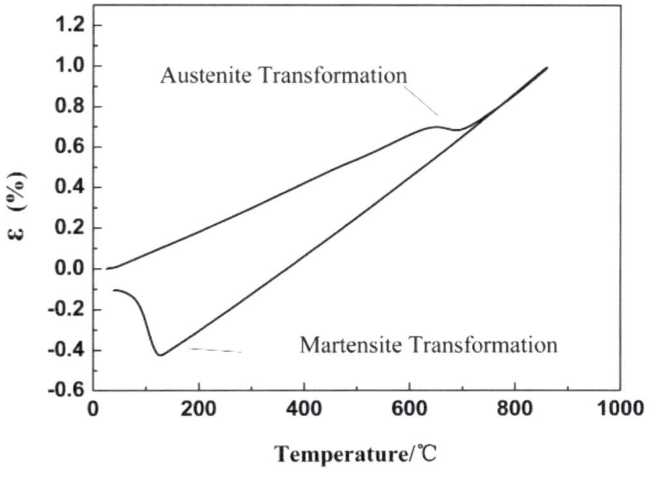

Figure 3. The free dilatometry test of FV520B steel.

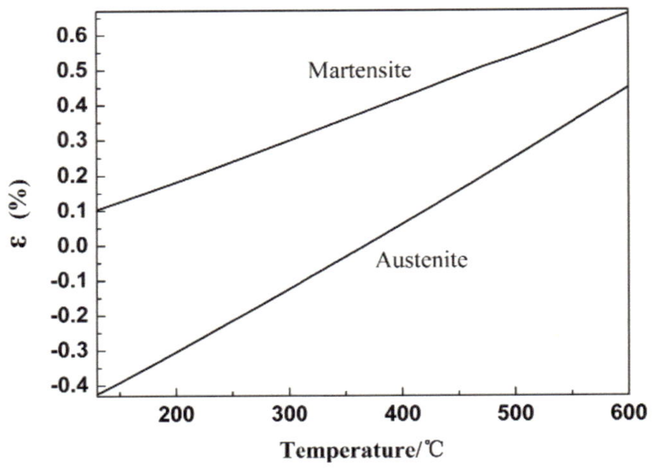

Figure 4. Temperature range of two phase regions during the free dilatometry test.

For materials in the temperature between M_s and A_c line, it could be martensite, austenite or dual-phase structure. The lower was the temperature, the higher was the density contrast of the two phase states. Consequently, the volume effect during phase transformation was more significant and the volume changes increased as the temperature drop. For the solid state phase transformation of the common steel, the transformation temperature of pearlite was the highest, next was bainite and that of martensite was the lowest [9]. In general, the volume effect of these phase transformations were corresponding to the transformation temperatures.

It had been noted above that superhigh heating rate and cooling rate had magnificent impact on the physical parameters of material during the laser remanufacturing. Studies indicates that the heating rate could reached over 500°C by methods of computer numerical simulation

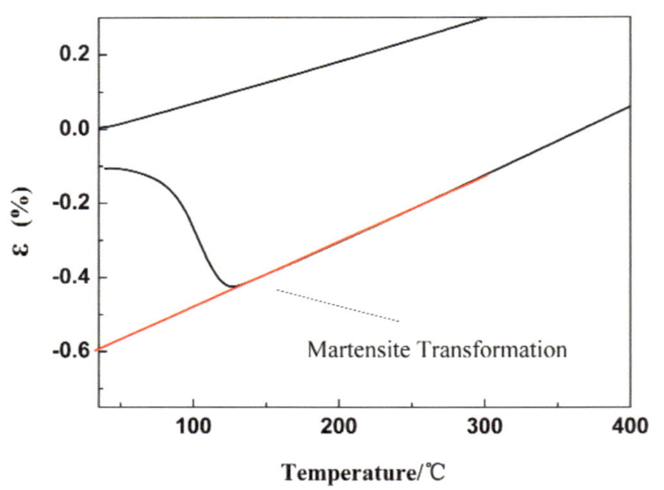

Figure 5. Temperature range of martensite transformation during the free dilatometry test.

http://dx.doi.org/10.5772/intechopen.79910

and temperature measurements. The simulation experiments were designed and conducted to study the effect of heating rate on the solid state phase transformation. Limited to the instrument condition, the heating rates were set as 0.25, 1, 5, 10, 30, 80 and 100°C/s. The results revealed that the rise of heating rates of each thermal cycles significantly increased Ac1 and Ac3 of FV520B (as shown in **Figure 6**). The expansion coefficient of martensite at 640°C increased apparently as the heating rate increased from 5 to 30°C/s. In general, heating rates had a great effect on the kinetics parameters of the austenite phase transformation.

2.2. The thermal cycles of laser cladding FV520B considering the effect of solid state phase transformation

In recent years, many studies were concerned about the effects of solid state phase transformation on the temperature and stress field and the mechanisms were revealed by experimental methods. However, the theoretical analysis were not enough in conditions of the complex projects like laser remanufacturing, which involved process control with many parameters. Therefore, computer numerical simulation was considered as a practical method to obtain precise, comprehensive and quantitative results, which had engineering meanings.

All the involved parameters were listed below: laser power, scanning rate, laser spot size and lapping rate were 1.5KW, 10 mm/s, 3 mm, 50% respectively. Using the technique of deactivate and reactivate element, the cladding material were activated by passes and layers. By building the coupling thermo-mechanical model, the evolution of the temperature field during the cladding was revealed. This work focused on the short and varying thermal cycles in positions with different depth in HAZ during the single-pass and multi-layer claddings. **Figure 7** was the geometries of the laser cladding samples during the single-pass and multi-layer processes [12, 14, 15].

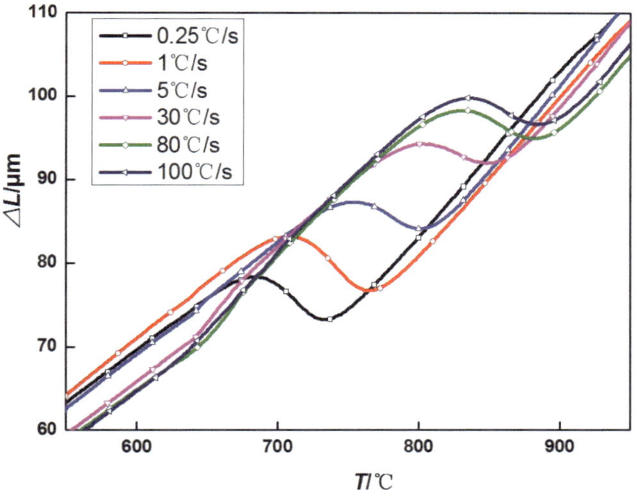

Figure 6. The free dilatometry test of FV520B steel in different level of heating rate.

Figure 8 was the temperature evolution at midpoint of single-pass cladding sample. The temperature of the melting pool reached 1833°C after 3.5 s. Meanwhile, the temperature measurement experiment at the midpoint was conducted by using RaytekMM infrared thermometer. The results showed that the laser beam scan midpoint of the single track after 3.7 s and the temperature of the melting pool was 2175°C, which was consistent with the simulation results.

Figure 9 was the thermal cycles in HAZ during the single-pass cladding. The maximum temperature in regions close to surface of the work-piece reached over 1000°C, which is higher than the complete austenitic temperature Ac3. In addition, the measurement results of the heating rate was around 500°C/s and that of the cooling rate was over 100°C. For regions in different depth of HAZ, the maximum temperatures of the thermal cycles were the main difference, which could directly affected the final microstructure in HAZ. Meanwhile, the simulation results could be provided as evidence for the studies on the mechanism of the microstructure evolution.

Figure 10 was the thermal cycles in HAZ during the multi-layer cladding. In multi-layer laser remanufacturing process, the coating is deposited layer by layer and HAZ was created repeatedly [16]. In addition, the repeatedly thermal effect on microstructure in HAZ was hard to control and predict. Some studies pointed that the microstructure in HAZ experienced approximate tempering during the repeatedly thermal cycles and toughening effect on microstructure was observed. The numerical simulation results showed that microstructure of HAZ experienced thermal cycles with over 600°C maximum temperature during the subsequent deposition process after the first layer cladding. It was considered that the multi-cycle heating and cooling process could affect the final microstructure and mechanical properties in HAZ.

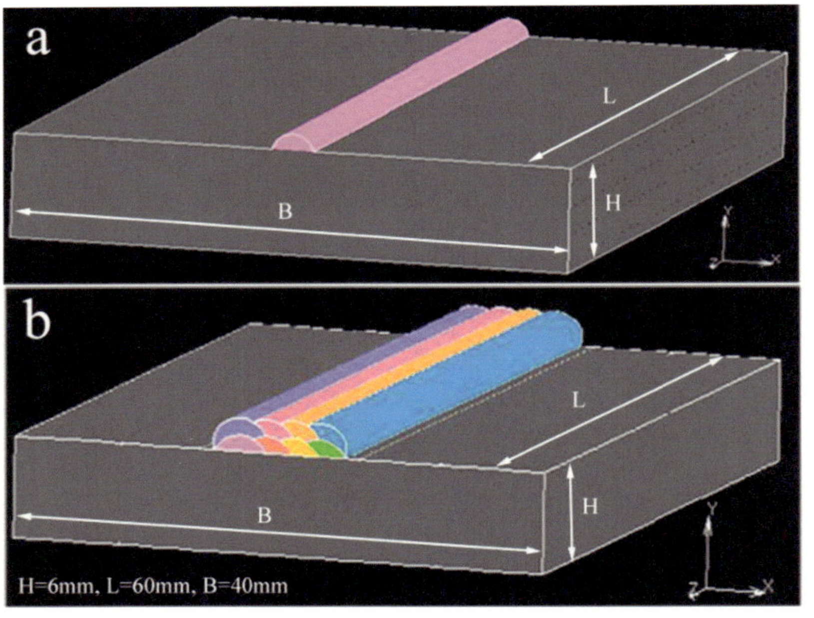

Figure 7. Geometry of laser clad sample: (a) single-layer; (b) multilayer.

2.3. The continuous cooling transformation (CCT) curve of FV520B

Figure 11 was the schematic of phase transformation for the martensite stainless steel during

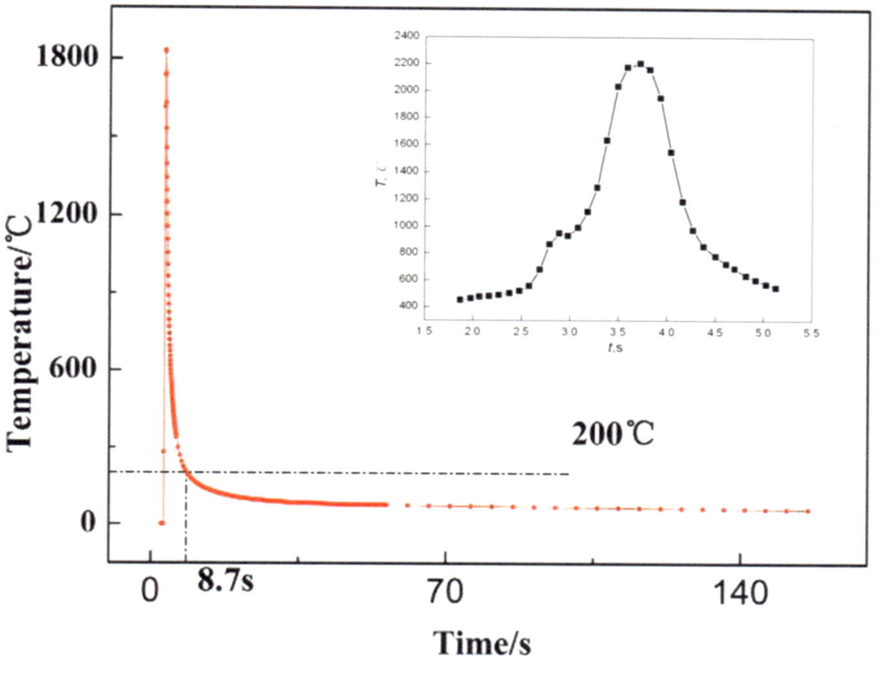

Figure 8. Temperature variation at midpoint of single-layer laser clad sample.

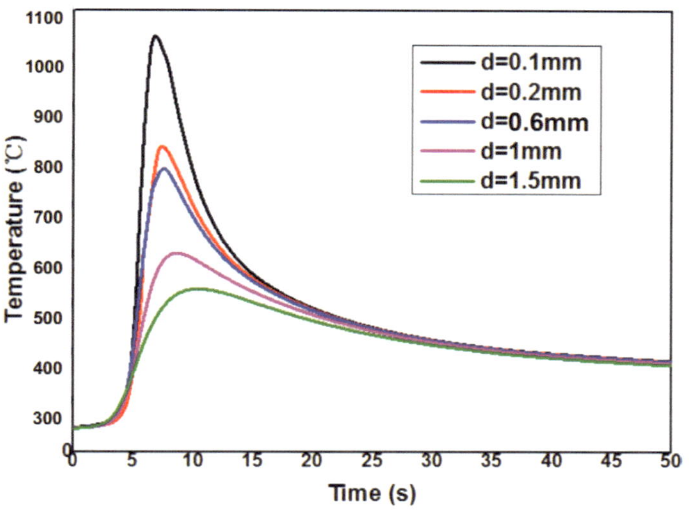

Figure 9. The thermal cycles in HAZ.

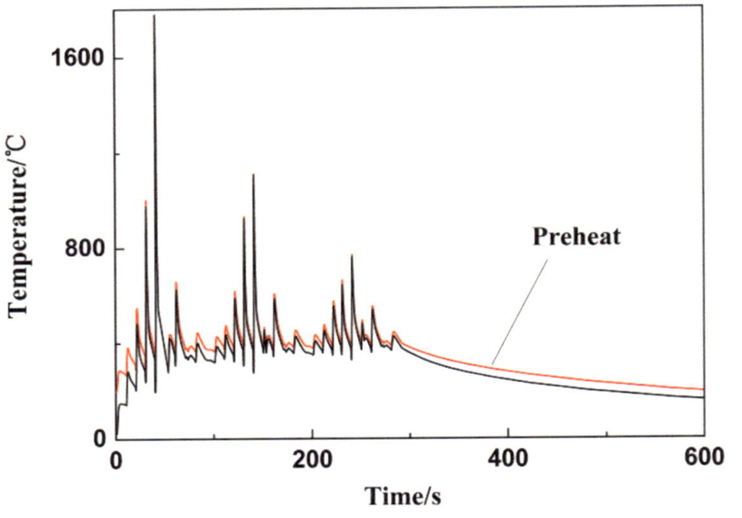

Figure 10. The temperature at the midpoint of the five pass of the first layer, with and without preheating.

the laser remanufacturing process. Austenite was assumed to be the initial phase during solidification process. In addition, the martensite transformation occurred during the cooling process due to the high cooling rates. As temperature decreased, the martensite transformation started at Ms and finished at Mf. When a new layer of FV520B was deposited, the previously deposited material experienced a new thermal cycle. Once the rising temperature was higher than Ac1, the transformation of martensite to austenite occurred. The percentage of austenite phase increased linearly as temperature rise.

As the physical properties of material depended on temperature and phases, the accuracy of the physical properties based on the appropriate description on the evolution of phase. In laser cladding process, different regions had different thermal cycles and external load, which

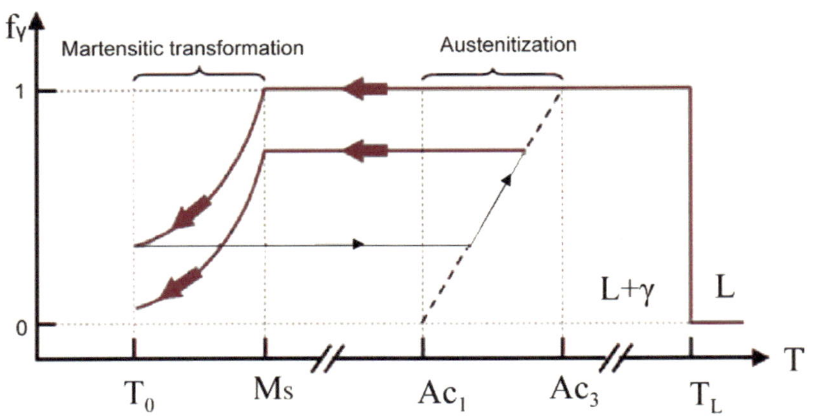

Figure 11. Evolution of volume fraction of austenite during phase transformation during Laser cladding directed shaping.

could contribute to the change of the kinetics and effect of solid state phase transformation. Therefore it was necessary to carry out systematical research in the influence of cooling condition on the solid state phase transformation during the cooling process [17].

A combined experiment by methods of dilatometry and metallography was conducted and the CCT curve was obtained by the L78 RITA transformation measuring apparatus. According to the thermal cycles of the laser cladding processes, two austenitic procedures were designed to study the effect of the austenitization condition on the microstructure evolution. The process of case 1 was designed as below: the sample heated up to 1000°C at heating rate of 300°C/s and kept the temperature for 5 s. In the cooling period, the sample was cooled to 810°C at the cooling rate of 100°C/s, kept the temperature for 5 mins and then was cooled at cooling rate of 0.02, 0.08, 0.2, 0.5, 1 and 50°C/s respectively. For case 2, the sample heated up to 1300°C at the same heating rate, was kept the temperature for 5 s and cooled to 810°C at the same cooling rate. Then, the sample was cooled at cooling rate of 0.02, 0.08, 0.2, 0.5, 1°C/s respectively. The phase transformation temperatures were obtained using tangent method.

Figure 12 was the continuous cooling transformation of FV520B stainless steel. **Figures 13** and **14** were metallograph of microstructures in different austenitic procedures. The results showed that the microstructures were all martensite in the designed cooling parameters due to the hardenability of FV520B stainless steel. For case 1, the microstructure was lath martensite with finely distributed precipitates. In addition, the proportion of martensite with sparse laths increased with the increasing cooling rate in the process. For case 2, the microstructure was lath martensite as well. However, the microstructure of martensite in case 2 was coarser than that of case 1 and dispersedly distributed precipitates were barely observed. As shown in **Figures 15** and **16**, the phase transformation temperature of case 2 was 20°C lower than that of case 1 and both increased as the cooling rate became higher; the microhardness of case 2 was 60 HV higher than that of case 1 and both decreased as the increasing cooling rate. In general, the maximum temperature of thermal cycle had a dominating effect on the microstructure, microhardness and phase transformation temperature rather than cooling rate.

It was considered that the maximum temperature was 1000°C with short heat preservation in case 1 and the sample experienced complete austenitic process with precipitates partly

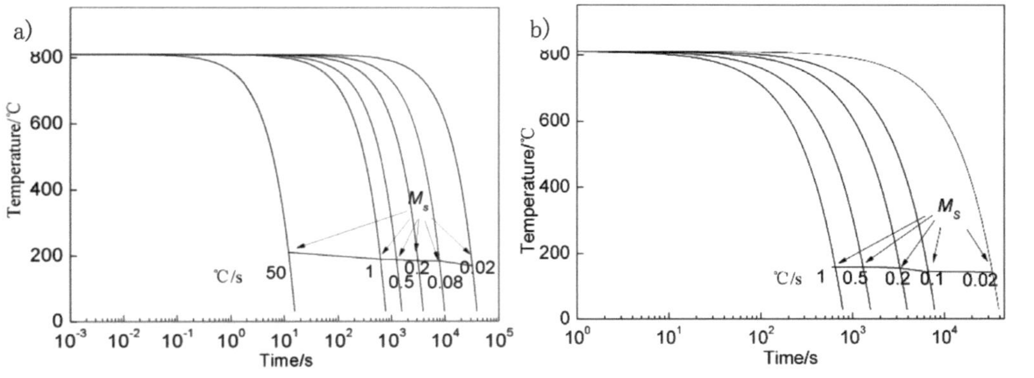

Figure 12. Welding continuous cooling transformation of FV520B stainless steel: (a) Case I; (b) Case II.

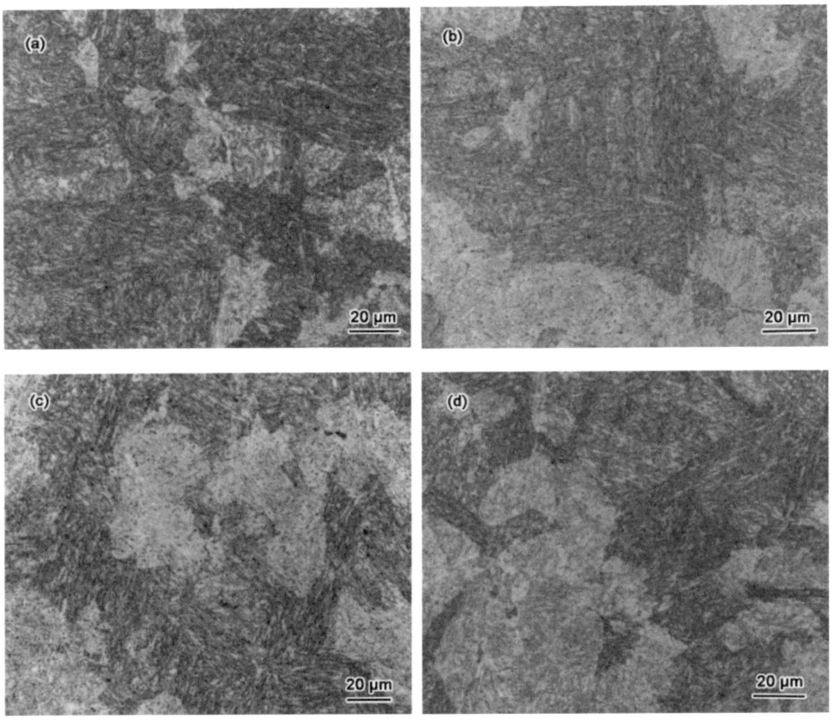

Figure 13. OM images of specimen Case I under different cooling rates. (a) 0.02°C/s; (b) 0.2°C/s; (c) 0.5°C/s; (d) 1°C/s.

dissolving in the matrix. In addition, the precipitates further grew coarse and solubility in the matrix decreased, which eventually contributed to the decreasing microhardness. However, the maximum temperature of case 2 reached 1250°C and no ferrite existed at high temperature. In general, the sample of case 2 experienced complete austenitic process with precipitates completely dissolving in the matrix. The increasing solid solubility of alloy elements contributed to the higher microhardness than that of case 1. Meanwhile, phase transformation temperature of case 1 and case 2 were both low and it could caused by the strengthen effect of austenite and growth of precipitates during the cooling period at low cooling rate.

The microhardness of laser remanufacturing specimens was higher than that from CCT process due to the superhigh heating rate in the actual laser cladding process. According to the numerical simulation results, the temperature gradient in high temperature period reached to 1000°C/s during the laser cladding process and the maximum heating rate could be set as 500°C/s due to the restrictions of instrument condition **Figures 15** and **16**.

As above, austenitic process and cooling rate had a remarkable impact on the phase transformation temperature and it could be interpreted as the effect of thermal cycle on the valid elements of material composition. It was assumed that the maximum temperature and heating rate during the heating period in the thermal cycle had great influence on the martensitic phase transformation temperature. Moreover, the strengthen effect and reducing of Ms took

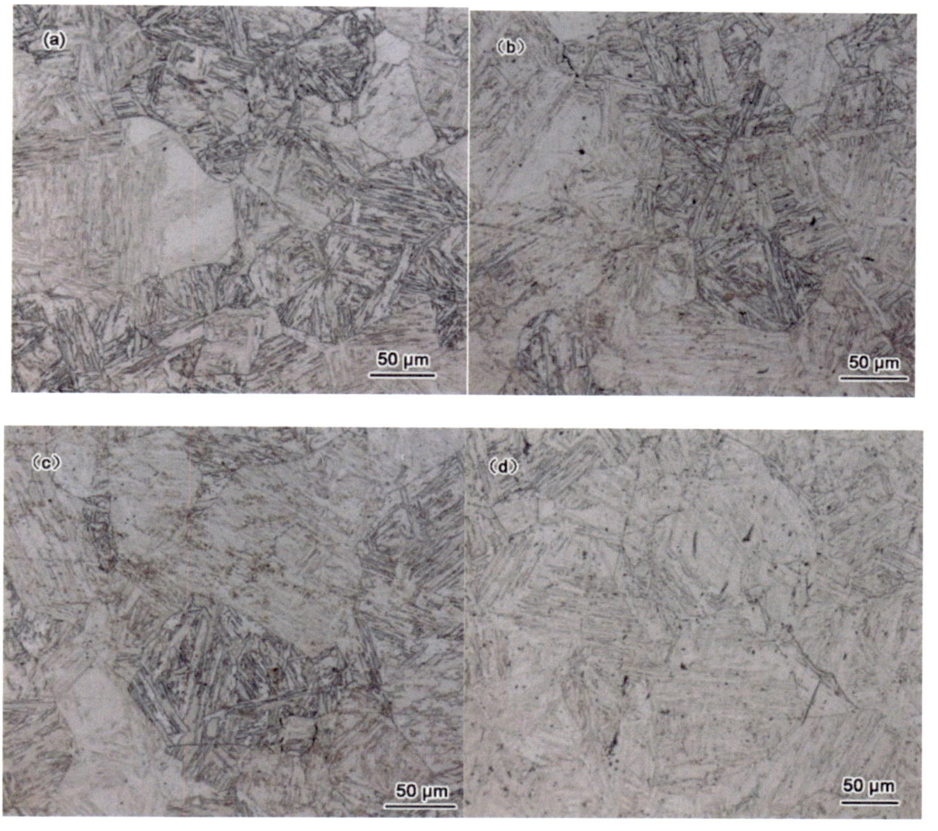

Figure 14. OM images of specimen Case II under different cooling rates. (a) 0.02°C/s; (b) 0.2°C/s; (c) 0.5°C/s; (d) 1°C/s.

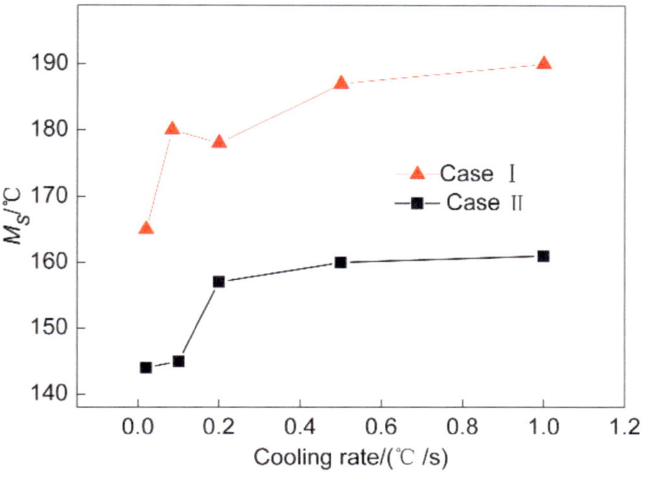

Figure 15. Martensite transformation start temperature under Case I and Case II.

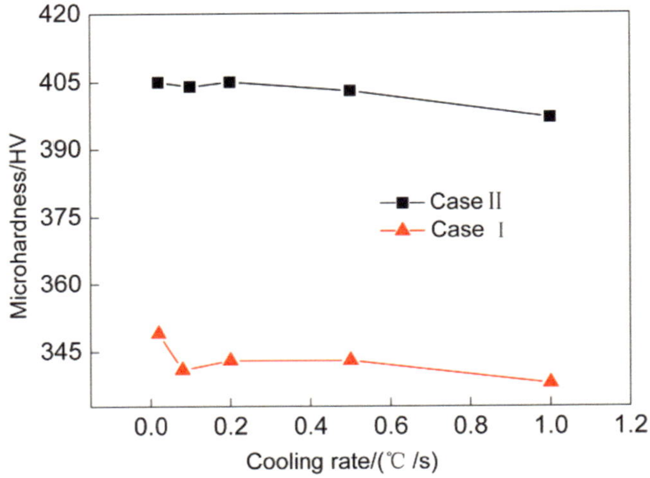

Figure 16. Hardness under the Case I and Case II.

place due to the low cooling rate during the same austenitic process. In general, the thermal cycles had a significant effect on the metallographic transformation [18, 19].

3. The Microstructure characteristics in HAZ of laser remanufacturing FV520B

Figure 17 were microstructure distribution of HAZ. It could be seen that the HAZ microstructure showed a stepwise characteristic. The microstructure of regions adjacent to interface was coarse martensite and no ferrite existed in high temperature. In thermal cycles of these regions, it experienced complete austenitic phase transformation and precipitates fully dissolved in the matrix and consequently the final microstructure of the regions were martensite with less lath characteristics. In the middle area of HAZ where the maximum temperature of the thermal cycle was close to Ac3 and the matrix underwent the complete austenitic as well. For the partially austenization zone, the laser energy input was relatively low and the matrix experienced partially austenization due to a thermal cycle with a maximum temperature between Ac1 and Ac3. Therefore, the microstructure was lath martensite with more dispersedly distributed precipitate and particles. The microstructure of regions in the bottom of HAZ was similar to the original microstructure of FV520B and it was considered to undergo the thermal cycle with a maximum temperature near Ac1.

Figure 18 was the microhardness distribution of cladding layer, HAZ and substrate. The microhardness distribution was considered consistent with that of microstructure. In the complete melting zone where the maximum temperature of thermal cycle was higher than AC3, it showed high hardness due to the solution strengthen. In the partially austenization zone, the microhardness was lower than other regions.

http://dx.doi.org/10.5772/intechopen.79910

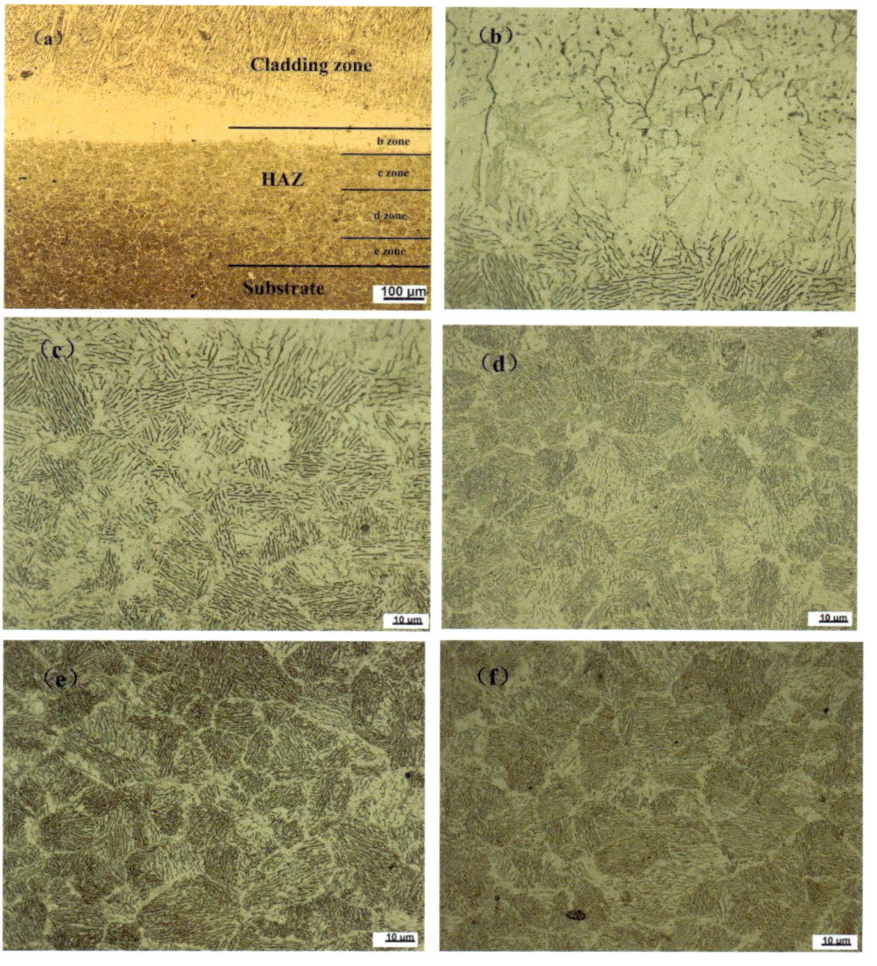

Figure 17. OM images of laser RM FV520B: (a) Macrostructure of laser RM FV520B (b) region close to the interface (c) partially molten zone (d) completely austenization zone (e) partially austenization zone (f) FV520B substrate.

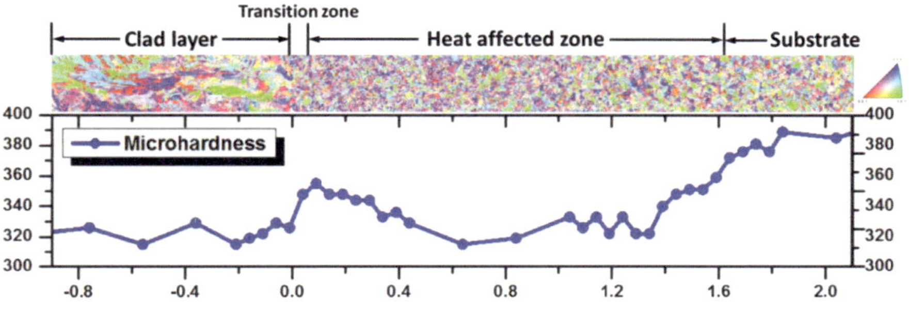

Figure 18. Microhardness distribution of HAZ-substrate.

4. Mechanical properties of laser remanufacturing FV520B

Figure 19 was the tensile test results of FV520B steel and cladding layer of laser remanufacturing FV520B. It could be seen that the yield stress, tensile stress and elongation of FV520B steel were respectively 830 Mpa, 970 Mpa and 23% and that of FV520B laser cladding layer were 920 Mpa, 1280 Mpa and 11% respectively. The results indicated that the strength of alloy increased and ductility decreased after laser remanufacturing process.

Figure 20 was the tensile test curves of laser remanufacturing butt samples and FV520B steel samples. The results showed that two samples had similar stress–strain curve. It could

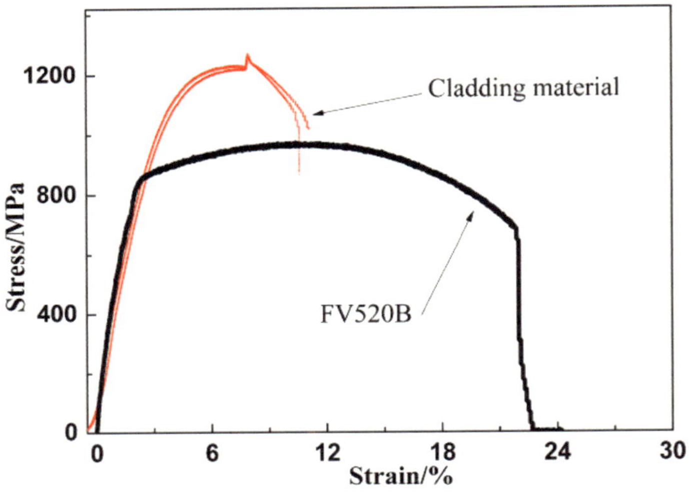

Figure 19. Comparison of tensile test curves between FV520B steel and as-deposited component.

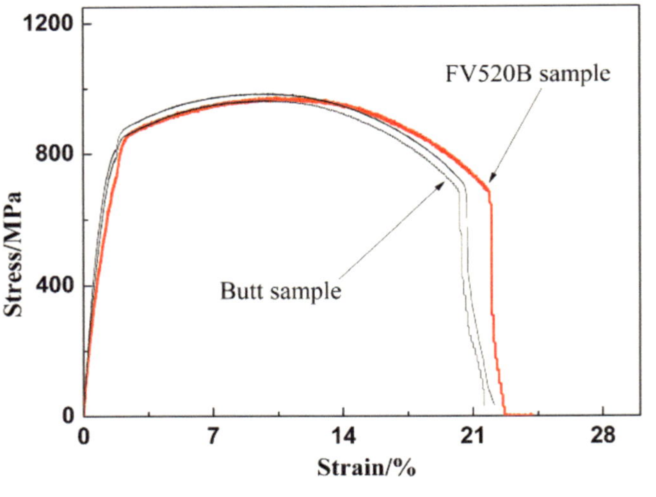

Figure 20. Comparison of tensile test curves between FV520B steel and butt sample.

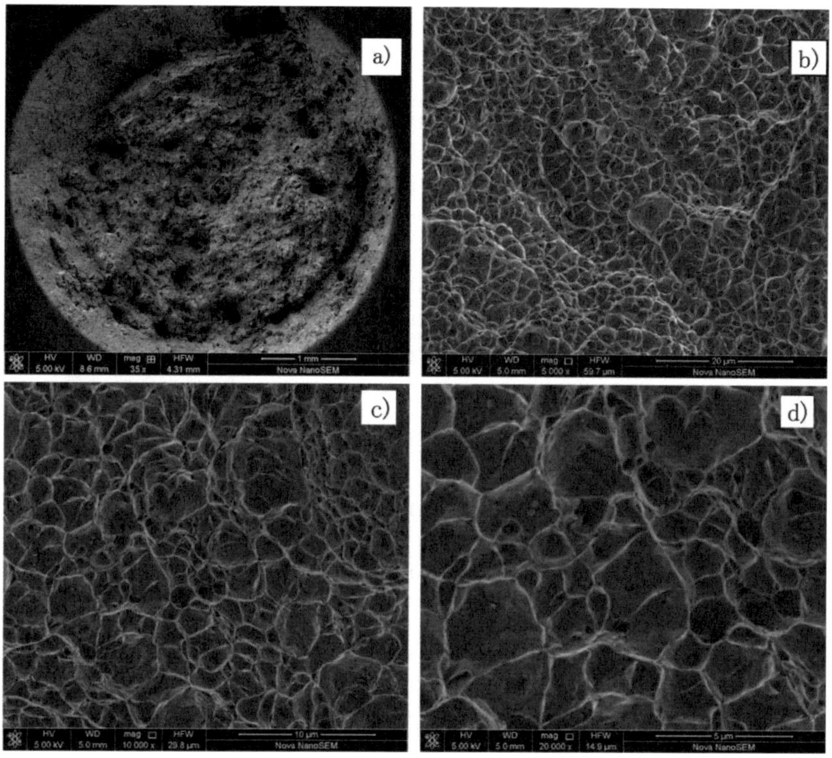

Figure 21. Fracture surface morphologies of FV520B steel fabricated by laser cladding. (a) Macro morphology of fracture surface; (b) fracture surface by 5000 times; (c) fracture surface by 10,000 times; (d) fracture surface by 20,000 times.

be predicted that the fracture section of butt samples was the bottom of HAZ due to the strengthen and descend of ductility of HAZ after laser remanufacturing process. In this research, FV520B work-piece underwent aging treatment after forging process and showed relative low strength but high ductility. In general, laser remanufacturing FV520B had high strength and ductility, which reached the combination properties of forgings.

Figure 20 was the fracture surface morphologies of laser remanufacturing FV520B samples. Ductile characteristics were observed without any apparent flaws or inclusion. In addition, High density and tiny dimples were observed at high magnification, where nanoscale spherical particles were founded. In general, the results showed that the material had good ductility with uniform distribution of nanoscale particles and no segregation at grain boundaries **Figure 21**, crack or inclusion were founded in the fracture surface [20].

5. Conclusions

Laser remanufacturing is an advanced repairing method to remanufacture damaged parts based on laser cladding processing. To reveal the mechanism of solid state phase transformation and the microstructure evolution during the laser cladding processing, it is necessary to study the effect of solid state phase transformation on physical properties and temperature filed by numerical simulation and experimental method. For FV520B stainless steel, the result of free dilatometry test showed that heating rates had a great effect on the kinetics parameters of the austenite phase transformation and the volume effect of phase transformations were corresponding to the phase transformation temperatures. According to the results of thermal simulation experiment and metallographic observation, it was considered that the multi-cycle heating and cooling process affected the final microstructure and mechanical properties in HAZ. The maximum temperature of thermal cycle had a dominating effect on the microstructure, microhardness and phase transformation temperature rather than cooling rate. Thermal cycle influenced significantly the metallographic transformation and consequently decided the final mechanical performance. By using the optimal process, laser remanufacturing FV520B had high strength and ductility, which reached the combination properties of forgings.

Acknowledgements

The work was supported by the National Key Research and Development of China (Grant No. 2016YFB1100205), and 973 Project of China (Grant No. 2011CB013403).

Author details

Shi-yun Dong*, Xiang-yi Feng, Jin-xiang Fang and Shi-xing Yan

*Address all correspondence to: syd422@sohu.com

National Key Laboratory for Remanufacturing, Academy of Armored Force Engineering, Beijing, China

References

[1] Wang HM. Materials' fundamental issues of laser additive manufacturing for high-performance large metallic components. Acta Aeronautica at Astronautica Sinica, China. 2014;**35**(10):2690-2693. DOI: 10.7527/S1000-6893.2014.0174

[2] Xu BS. Theory and Technology of Equipment Remanufacturing Engineering. Beijing, China: National Defense Industry Press; 2007

[3] Walsh B. PSS for product life extension through remanufacturing. CIRP IPS2 Conference, Sweden: Linköping University Electronic Press; 2010:261-266

[4] Xu BS, Dong SY. Laser Remanufacturing Technology. Beijing, China: National Defense Industry Press; 2016

[5] Lund RT. The Remanufacturing Industry: Hidden Giant. Boston, Massachusetts: Boston University; 1996

[6] Nasr N, Thurston M. Remanufacturing: A Key Enabler to Sustainable Product Systems. Rochester, NY-USA: Rochester Institute of Technology; 2006

[7] Qiu CL, Ravi GA, Dance C, et al. Fabrication of large Ti-6Al-4V structure by direct laser deposition. Journal of Alloys and Compounds. 2015;**629**:351-361. DOI: DOI 10.1016/j.jallcom.2014.12.234

[8] Mower TM, Long MJ. Mechanical behavior of additive manufactured, powder-bed laser-fused materials. Materials Science and Engineering A. 2016;**651**:198-213. DOI: 10.1016/j.msea.2015.10.068

[9] Yadollahi A, Shamsaei N, Thompson SM. Effects of process time interval and heat treatment on the mechanical and microstructural properties of direct laser deposited 316L stainless steel. Materials Science and Engineering. 2015;**644**:171-178. DOI: 10.1016/j.msea.2015.07.056

[10] Wang T, Zhu YY, Zhang SQ, et al. Grain morphology evolution behaviour of titanium alloy components during laser melting deposition additive manufacturing. Journal of Alloys and Compounds. 2015;**632**:505-513. DOI: 10.1016/j.jallcom.2015.01.256

[11] Goldak JA, Akhlaghi M. Computational Welding Mechanics [M]. New York: Springer US; 2006

[12] Costa L, Vilar R, Reti T, et al. Rapid tooling by laser powder deposition: Process simulation using finite element analysis. Acta Materialia. 2005;**53**(14):3987-3999. DOI: 10.1016/j.actamat.2005.05.003

[13] Xu ZY. Martensitic Transformation and Martensite. Beijing, China: Science Press; 1999

[14] Francis JA, Bhadeshia H, Withers PJ. Welding residual stresses in ferritic power plant steels. Materials Science and Technology. 2007;**23**(9):1009-1020. DOI: 10.1179/174328407X213116

[15] Börjesson L, Lindgren LE. Simulation of multi-pass welding with simultaneous computation of material properties. Journal of Engineering Materials and Technology. 2001;**123**(1):106-111

[16] Chen Y, Zhang K, Huang J, et al. Characterization of heat affected zone liquation cracking in laser additive manufacturing of Inconel 718. Materials and Design. 2016;**90**:586-589. DOI: DOI 10.1016/j.matdes.2016.10.155

[17] Koistinen DP, Marburger RE. A general equation prescribing the extent of the austenite-martensite transformation in pure iron-carbon alloys and plain carbon steels. Acta Metallurgica. 1959;**7**(1):59-60

[18] Fang JX, Dong SY, Wang YJ, et al. The effects of solid-state phase transformation upon stress evolution in laser metal powder deposition. Material and Design. 2015;**87**:807-814. DOI: 10.1016/j.matdes.2015.08.061

[19] Xu BS, Fang JX, Dong SY, et al. Heat-affected zone microstructure evolution and its effects on mechanical properties for laser cladding FV520B stainless steel. Acta Metallurgica Sinica. 2016;**52**(1):1-6. DOI: 10.11900/0412.1961.2015.00489

[20] Fang JX. Effects of solid-state phase transformation upon stress evolution during laser cladding forming of martensitic stainless steel. Harbin Institute of Technology; 2016

Chapter 5

Duplex Stainless Steels: Effect of Reversion Heat Treatment

V. Shamanth, K. S. Ravishankar and K. Hemanth

Additional information is available at the end of the chapter

http://dx.doi.org/10.5772/intechopen.80007

Abstract

Duplex stainless steels present good corrosion resistant and mechanical properties hence they are being used in various pressure boundary components of nuclear power plants such as primary coolant pipes, valves and pump bodies because of the presence of the dual microstructure which consists of equal amounts of austenite and α-ferrite phases. The ratio of austenite and α-ferrite phases mainly depends on the chemical composition. However, when DSS are subjected to the service temperature range of about 300–500°C they undergo embrittlement due to spinodal decomposition of highly alloyed ferrite matrix into iron rich (α) phase and chromium rich (α') phase. The embrittlement significantly affects impact toughness, tensile strength, ductility, fracture toughness and fatigue behavior limiting the industrial applicability of this steel to temperatures below 280°C. In this chapter, the basic overview of duplex stainless steels and the effect of reversion heat treatment on the thermally embrittled duplex steel is discussed.

Keywords: duplex stainless steels, 475°C embrittlement, reversion heat treatment, alpha and alpha-prime precipitates

1. Introduction

Stainless steels are alloys of iron containing at least 11% chromium by weight. They exhibit superior corrosion resistance compared to other steels mainly due to the passive film of chromium oxide which forms on the surface. However, in order for a stainless steel to retain its "stainless-ness" in aggressive chemical environments, larger amounts of chromium need to be added to the alloy [1]. In addition to chromium, other alloying elements such as nickel, molybdenum, manganese, nitrogen etc. are also added in order to provide better resistance to different forms of corrosion. Some alloying elements are also added to enhance mechanical properties and weldability without compromising on the corrosion resistance [2]. Stainless steels can be classified into five groups based on their microstructures: ferritic, austenitic, martensitic, duplex and precipitation hardening stainless steels [3].

Duplex stainless steels (DSS) have a microstructure consisting of ferrite and austenite in nearly equal proportions and exhibit better corrosion resistance and mechanical properties in comparison to single phase stainless steels [4]. They provide excellent resistance to pitting corrosion and stress corrosion cracking even in chloride environments. DSS can be classified into three families based on their pitting resistance equivalent numbers (PREN). They are: "lean duplex" alloys having a PREN slightly higher than that of austenitic stainless steel grades and contain up to 20 wt% Cr and no molybdenum, "standard duplex" alloys with around 22 wt% Cr and 3 wt% Mo, having a PREN lying between 33 and 36, and, "superduplex" alloys with more than 25 wt% Cr, 3.5 wt% Mo and 0.2–0.3 wt% N, with a PREN greater than 40. Recently, special grades called "hyperduplex" stainless steels with much higher chromium and nitrogen levels have also been developed [5].

DSS find applications in various industries which involve hostile environments and such as chemical and petrochemical, oil and gas, pulp and paper, power generation, hydrometallurgy, marine transportation, construction etc. which involve hostile environments. A major limiting factor in the applicability of DSS is their susceptibility to thermal embrittlement when exposed to temperatures in the range of 280–525°C. This form of embrittlement is popularly known as "475°C embrittlement" since the rate of embrittlement within this temperature range is maximum at 475°C [6]. The embrittlement is believed to be because of the formation of Cr-rich precipitates called "α'" in the ferrite phase. The embrittlement is accompanied by a drop-in corrosion resistance because of the depletion of chromium levels within the ferrite phase and around the precipitates [7].

Studies conducted on ferritic and duplex stainless steels have shown that the changes in microstructure and mechanical properties associated with the 475°C embrittlement can be undone by subjecting the material to a short term "reversion" heat treatment within the range of 550–600°C [8]. Although the effect of subjecting a 475°C embrittled DSS to reversion heat treatment has been studied by investigators in the past. In our investigation the focus was mainly on the recovery of microstructure and mechanical properties. However, when cast austenitic stainless steels are exposed to the 550°C or higher for a long time, various types of detrimental intermetallic phases, such as sigma (σ) phase, Chi (χ) phase, R phase, and M23C6 carbides could be formed in the ferrite phase or at ferrite phase boundaries. As a result, the mechanical properties will be significantly degraded. Therefore, to apply the reversion heat treatment to alleviate the thermal aging embrittlement, the possibility of inadvertent embrittlement of DSS by the reversion heat treatment has been carefully investigated. However, it was not completely clear that how these microstructural changes translate on mechanical properties of DSS and vice-versa. In addition, there was a concern that the recovered DSS after the reversion heat treatment could be re-embrittled faster when it is subjected to the service temperature of nuclear power plants [9]. Therefore, understanding the re-aging behaviors of the recovered DSS is also important to determine the applicability of the reversion heat treatment. In this regard, the Scope of present work was aimed to develop an economic approach to extend the service life of long term working duplex stainless steel components which are susceptible for embrittlement.

2. Duplex stainless steels: development and applications

The development of stainless steels began in the early twentieth century in the United Kingdom and Germany. The austenitic Fe-Cr-Ni steels became the largest group of stainless steels even though the earliest grades were martensitic and ferritic Fe-Cr steels. The minimum carbon levels in these steels were high (around 0.08%), due to existing furnaces and refining techniques, making them prone to grain boundary carbide precipitation during heat treatment and welding, and sensitive to intergranular corrosion attack.

The first reference to Fe-Cr-Ni duplex stainless steels with between 30 and 70% ferrite was made by Bain and Griffith in 1927. Soon after, several countries explored such steels in cast form. The first commercial duplex stainless steel product, a grade called 453E, was made in 1929 with an approximate composition of 25%Cr–5%Ni followed by a modified grade, 453S, with 25%Cr–5%Ni–1%Mo which was marketed in 1932. Applications in the form of castings involved autoclaves for gunpowder production and valves for sulfide pulping and as coolers of the Brobeck type in the form of plate and forgings. By the late 1940s, duplex stainless steels with high volume fractions of ferrite in an austenitic matrix, which were not sensitive to intergranular corrosion (IGC) even in the most aggressive media were available and marketed simultaneously in France, Sweden and the USA.

Development of low nickel-content duplex alloys was encouraged due to nickel shortage following the Korean War in 1950–1951, and it became apparent that a balance of ferrite and austenite provided better resistance to stress corrosion cracking (SCC) in chloride medium than fully austenitic grades. In France a grade with 20–35% ferrite called UR50 (UNS32404) was marketed in various forms for industries such as oil refining, food processing, pulp and paper, and pharmaceuticals. These steels were produced in high frequency induction furnaces with a partial vacuum that ensured carbon removal, rudimentary de-oxidation and restricted nitrogen ingress.

The understanding of the physical metallurgy of DSS had not progressed sufficiently to offer a material having good ductility and toughness that was easy to manufacture and fabricate. This gave them a reputation for crack sensitivity until the 1960s. During the late 1960s and early 1970s, there was another nickel shortage and pushed the price of austenitic alloys. With the introduction of vacuum and argon oxygen decarburization practices, steel production techniques had improved dramatically, leading to steels with low carbon, sulfur and oxygen contents along with greater control of nitrogen content. The maximum carbon content in the UNS S31500 was 0.03%. The chemical composition was controlled to optimize the ferrite-austenite phase balance and to enhance resistance to SCC. The addition of nitrogen improved corrosion resistance and stability of austenite in the heat affected zone (HAZ). However, they were still susceptible to IGC under certain conditions due to high Cr/Ni equivalent ratio, which led to an almost fully ferritic HAZ after welding. Carbides and nitrides precipitated along grain boundaries because of the very low solubility of carbon and nitrogen in ferrite. Higher levels of nickel and nitrogen were added to overcome this problem and the IGC resistance was significantly improved.

Through the 1970s and 1980s, several duplex grades were developed with emphasis on improved weldability and better resistance to corrosion in aggressive environments. Highly alloyed duplex grades such as the UNS S32550 with 25% Cr and up to 2% Cu, were developed as castings such as pumps and valves, and have been used extensively in the offshore industry and for sulfuric acid service. The superduplex grades containing about 25%Cr, 6–7%Ni, 3–4%MO, 0.2–0.3%N, 0–2%Cu and 0–2% W have pitting resistance equivalent numbers (PREN) greater than 40. In making superduplex alloys, care was taken to balance the Cr and Ni forming elements and higher levels of nitrogen were added. These factors stabilized the HAZ during welding, but promoted intermetallic precipitation [10].

In recent years duplex grades have emerged as an alternative to austenitic grades such as 316 and 304. Lean DSS are used in bridges, storage tanks and also for construction of transport vehicles. Superduplex grades such as Zeron 100 (UNS S32760) were developed to compete with super-austenitic grades and are used in large quantities in umbilicals for the control of sub-sea systems. DSS have also replaced austenitic grades in flue gas cleaning systems and desalination plants .

In natural gas pre-heaters, S32304 are selected for exchanger tubing where low grade steam is used for heating purposes. For applications where higher chloride contents are present, tubing is made from S31803/S32205. This grade is also used for reactors, storage tanks and heat exchangers in the production of detergents comprising of fatty amines and chlorides, in plastic production, in steam sterilization of bi-products of sodium cyanide production and so on. Cast 32550 is used in phosphoric and sulfuric acid production and also as stud bolts in ammonia injectors and valve internals in urea recycle lines.

Grade S31803/S32205 has been extensively used for the construction of advanced chemical tankers for marine transportation since 1987. Duplex grades are also used as propeller shafts, thrusters, water jet engines and other components subjected to high mechanical loads. Superduplex grades such as S32760 are used in pumps to handle potash at around 80°C and hot bauxite slurry in the Bayer process. Other uses for this grade include pipe work in a titanium dioxide refinery for spent hydrochloric acid lines. Low alloy grades such as S32304 are selected for water heaters, calorifiers and hot water tanks in breweries and similar industries. The cast DSS of the CF series finds applications in nuclear power plants for reactor coolant and auxiliary system piping [10].

3. Metallurgy of duplex stainless steels

During solidification of the DSS melt, the first solid that forms is δ-ferrite. As the temperature drops, austenite formation takes place. After complete solidification, the microstructure is that of austenite islands in a matrix of ferrite. The volume fraction of ferrite-austenite depends on the chemical composition. The large amounts of alloying elements added to DSS results in the formation of various carbides, intermetallics and other secondary phases which form over different temperature ranges at varying rates (**Figure 1**).

3.1. Secondary phases

3.1.1. Sigma phase

The Sigma (ζ) phase is a Cr, Mo rich hard embrittling precipitate which forms between 650 and 1000°C is often associated with reduction in impact toughness and corrosion resistance. Since

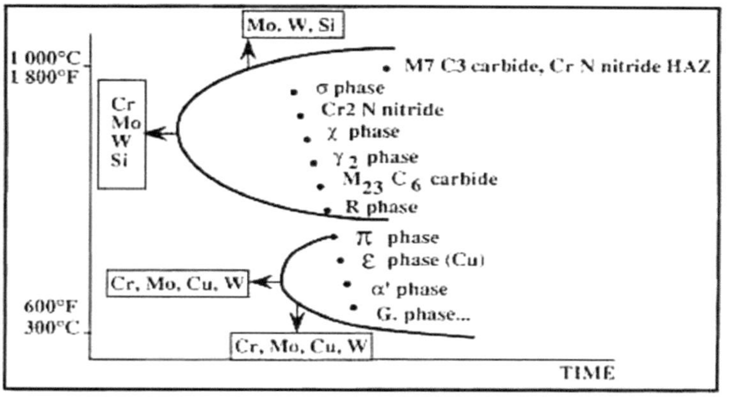

Figure 1. A schematic TTT curve for formation of precipitates in DSS and the effects of alloying elements in the temperature ranges of formation for various precipitates [10].

the mobility and concentration of Mo and Cr in ferrite is higher than in austenite, ζ-phase precipitation generally occurs in the ferrite phase. It also forms in the HAZ during welding. It has a tetragonal crystal structure with 32 atoms per unit cell and 5 different crystallographic atom sites. The morphology of ζ-phase changes with temperature. At around 750°C, it has a coral-like structure; at 950°C it is bigger and more compact [10].

The depletion in Mo content is a lot more pronounced compared to that of Cr, indicating that Mo is the main element controlling the precipitation of ζ-phase. The formation of ζ-phase is rapid and a very high cooling rate is required to avoid it formation during quenching from solutionizing temperature. For a 2205 DSS, a cooling rate of 0.23 K/s is necessary to avoid more than 1% ζ-phase formation [10].

3.1.2. Chi phase

The enrichment of ferrite with intermetallic forming elements during long term thermal exposure at temperatures around 700°C, favors the precipitation of Chi (χ) phase. It often nucleates at the δ/γ interface and grows into the δ matrix. It is difficult to study its influence on corrosion and toughness since it often co-exists with ζ-phase. Increase in aging time causes an enrichment of Mo and depletion of Fe in the χ-phase. During isothermal aging, the χ-phase always precipitates before ζ-phase, but during continuous cooling, χ-phase appears only at low cooling rates [11].

3.1.3. Secondary austenite

The mechanism and rate of formation of secondary austenite (γ_2) may vary depending on the temperature. In the temperature range of 700–900°C, typical mechanism is by the eutectoid reaction, which is facilitated by rapid diffusion along the δ/γ boundaries giving rise to ζ-phase and γ_2 in prior ferrite grains. This also reduces the Cr and Mo content in the ferrite. When Cr2N precipitates cooperatively, γ_2 has been found to be poor in Cr, making it highly susceptible to pitting corrosion. At temperatures above 650°C, at which diffusion rates are higher, γ_2 is formed as Widmanstätten precipitates. Below 650°C, ferrite transforms to γ_2 by a mechanism quite similar to that of martensite formation. The γ_2 formed in this manner has a similar composition to the ferritic lattice thus indicating that the transformation was diffusionless [12].

3.1.4. *R-phase*

Isothermal treatment of duplex stainless steels between 550 and 650°C results in the uniform and very fine distribution of R-phase throughout the δ grains [11]. The R-phase is a Mo rich intermetallic having a trigonal crystal structure. Its formation reduces the toughness and critical pitting temperature in DSS. R-phase precipitates may be intergranular or intragranular in nature; the former perhaps more deleterious with regard to pitting corrosion since they may contain up to 40% Mo [12]. With the increase in aging time, R-phase transforms into ζ-phase due to diffusion of Mo from the R-phase into the ζ-phase, which eventually results in the decrease in the volume fraction of R-phase [13].

3.1.5. *Chromium nitrides*

The solubility of nitrogen at about 1000°C in ferrite is high, but drops on cooling and the ferrite becomes supersaturated in nitrogen, leading to the intergranular precipitation of needle-like Cr_2N. Isothermal heat treatment in the temperature range of 700–900°C usually results in precipitation of Cr_2N either on the δ/δ grain boundaries or the δ/γ phase boundaries. The hexagonal Cr_2N formed under these conditions has a negative influence on pitting corrosion resistance. In HAZ of welds, however, the cubic Cr_2N is the predominant nitride that has been observed [12]. Cr_2N precipitates display film-like or tiny platelet-like morphology.

3.1.6. *Carbides ($M_{23}C_6$ and M_7C_3)*

M_7C_3 forms at the δ/γ grain boundaries in the temperature range of 950–1050°C but can be avoided by ordinary quenching methods since its formation takes at least 10 min. $M_{23}C_6$ precipitates rapidly between 650 and 950°C, predominantly at the δ/γ boundaries where Cr-rich ferrite intersects with C-rich austenite. Several precipitate morphologies have been recorded including cuboidal, acicular and cellular form; each having an associated Cr depleted zone in its vicinity. Since modern duplex grades contain less than 0.02%C, carbides of either form are rarely seen [10].

3.1.7. *Alpha prime*

Alpha Prime (α') is a Cr-rich precipitate that forms in the temperature range of 280–525°C [4]. The main cause for formation of α' is the miscibility gap in the Fe-Cr system whereby ferrite undergoes spinodal decomposition into Fe-rich δ-ferrite and Cr-rich α'. Within the miscibility gap but just outside the spinodal, classical nucleation and growth of α' occurs. The α' precipitate has a body-centered crystal structure and is the main cause of hardening and 475°C embrittlement in ferritic stainless steels [10].

3.1.8. *Epsilon phase*

In duplex alloys containing copper, the supersaturation of ferrite due to decrease in solubility at lower temperatures leads to the precipitation of extremely fine particles of Cu-rich Epsilon (ε) phase within the ferrite grains after 100 h at 500°C. This significantly extends the low temperature hardening range for duplex stainless steels. Often, ε-phase has been mistaken for γ2 due to similar temperature ranges of formation [10].

3.1.9. G, π and τ phases

The G-phase develops at α/α' interfaces between 300 and 400°C after several hours of exposure, due to enrichment of Ni and Si at these locations [10].

The π-nitride is a Cr and Mo rich precipitate with a cubic crystal structure that forms at intergranular sites in DSS welds after isothermal treatment at 600°C for several hours [10].

The η-phase is a heavily faulted precipitate with needle-like morphology that forms due to heat treatment in the temperature range of 550–650°C. It has an orthorhombic crystal structure [11].

4. Effect of alloying elements

4.1. Chromium

The primary role of chromium in stainless steels is to improve the localized corrosion resistance, by the formation of a passive Cr-rich oxy-hydroxide film. This film extends the passive range and reduces the rate of general corrosion. The beneficial effect of adding very high levels of chromium is, however, negated by the enhanced precipitation of intermetallic phases which often lead to a reduction in ductility, toughness and corrosion resistance. Apart from this chromium also stabilizes ferrite [10]. Although, other alloying elements can influence the effectiveness of the passive film, none of them can create the properties of stainless steel, by themselves. However, it is often more efficient to improve corrosion resistance by addition of other elements, with or without increasing the chromium content to ensure that the mechanical properties, fabricability, weldability or high temperature stability remain largely unaffected [14].

4.2. Nickel

Nickel, when added in sufficient quantities, stabilizes austenite; this greatly enhances mechanical properties and fabrication characteristics. Nickel effectively promotes re-passivation, especially in reducing environments and is particularly useful in resisting corrosion in mineral acids. Increasing nickel content to about 8–10% decreases resistance to SCC, but on further increase, SCC resistance is restored and is achieved in most service environments at about 30% Ni [14].

In order to maintain 40–60% ferrite, balance austenite in DSS, the ferrite stabilizing agents need to be balanced with the austenite stabilizers. For this reason, the level of nickel added to a DSS will depend primarily on the chromium content. Excessive nickel contents may enhance intermetallic precipitation when the alloy is exposed to the temperature range of 650–950°C, due to enrichment of ferrite in Cr and Mo. High Ni contents also accelerate α' formation. Although nickel does have some direct effect on corrosion properties, it appears that its main role is to control phase balance and element partitioning [10].

4.3. Molybdenum

Molybdenum, in combination with chromium, effectively stabilizes the passive film in the presence of chlorides. Molybdenum is effective in increasing the resistance to the initiation of pitting and crevice corrosion [14]. Its effect on ferrite stability is similar to that of chromium. To prevent ζ-phase formation in the hot working temperature range, i.e. above 1000°C, the upper limit of Mo addition of is limited to about 4%, since Mo is the main element controlling the precipitation of ζ-phase.

4.4. Manganese

Although manganese acts as an austenitic stabilizer in austenitic stainless steels, mixed results have been obtained for DSS in which it has little effect on the phase balance. It appears that Mn increases the temperature range and formation rate of ζ-phase. Manganese increases abrasion and wear resistance and tensile properties of stainless steels without loss of ductility. Further, Mn increases the solubility of nitrogen, thus allowing for higher nitrogen contents. However, Mn additions in excess of 3 and 6%, for nitrogen levels of 0.1 and 0.23% respectively, significantly decrease the critical pitting temperature (CPT). Nevertheless, the combination of Mn and N in modern DSS improves the pitting resistance and counteracts the singular problems associated with Mn [10].

4.5. Nitrogen

Nitrogen enhances pitting resistance by retarding the formation of ζ-phase and diminishes Cr and Mo segregation and also raises the corrosion resistance of the austenitic phase in DSS [14]. Nitrogen has also been reported to increase crevice corrosion resistance. Nitrogen strengthens austenite by dissolving at the interstitial sites in solid solution. Addition of nitrogen to DSS, suppresses austenite dissolution and encourages austenite reformation in the HAZ [10].

4.6. Copper

Copper reduces the corrosion rate of high alloy austenitic grades in non-oxidizing environments, such as sulfuric acid. In some DSS with 25% Cr, 1.5% Cu is added to obtain the optimum corrosion resistance in 70% H_2SO_4 at 60°C. For boiling HCl, an addition of 0.5% Cu decreases both active dissolution and crevice corrosion rates. Copper additions in DSS are limited to about 2%, since higher levels, reduce hot ductility and can lead to precipitation hardening. Exposure at temperatures between 300 and 600°C can lead to precipitation of tiny Cu-rich precipitates which do not significantly reduce corrosion resistance or toughness but can be exploited for improving abrasion-corrosion resistance in duplex pump castings [10].

4.7. Tungsten

Up to 2% tungsten additions have been made in DSS to improve pitting resistance. Tungsten also increases crevice corrosion resistance in heated chloride solutions. Tungsten encourages intermetallic formation in the 700–1000°C temperature range, and also encourages γ_2 formation in weld metal. Thermodynamically, it is believed to be equivalent to Mo with respect to ζ-phase formation, but this is not the case in terms of kinetics. Levels between 1 and 3% have been shown to restrict ζ-phase formation to the intergranular sites instead of phase boundaries; the influence of the large tungsten atom on the diffusion of Mo and W at the phase boundaries is thought to be the reason. Tungsten alloyed weld metal has been shown to form χ-phase more rapidly than in welds without W additions. Generally, the tungsten content is limited to 1% in a 4% Mo DSS, and 2% in DSS with about 3% [10].

4.8. Silicon

Silicon enhances high temperature oxidation resistance and is also beneficial for concentrated nitric acid service. DSS bearing high silicon (3.5–5.5%) have enhanced pitting corrosion resistance and a claimed immunity to SCC. However, it is preferred to limit Si additions to 1% since Si is generally considered to enhance ζ-phase formation [10].

4.9. Carbon, sulfur and phosphorous

The carbon content of most wrought DSS is limited to 0.02 to 0.03%, primarily to suppress precipitation of Cr-rich carbides along the grain boundaries. Sulfur and phosphorous contents are controlled but not eliminated. The presence of S is important for weld bead penetration. Modern steel making processes such as argon oxygen decarburization (AOD) and vacuum oxygen decarburization (VOD) help in controlling the levels of S and C, while P contents can be reduced by using good melting practice [10].

5. Effect of heat treatment

5.1. Solution annealing

Element solubility in ferrite falls with decreasing temperature, increasing the probability of precipitation during heat treatment. During solidification, DSS solidifies completely as ferrite and then undergo solid state transformation into austenite. This is a reversible process and as a result, any large increases in temperature above 1000°C lead to an increase in ferrite and also a reduction in the partitioning of substitutional elements between phases. In addition, ferrite becomes enriched in interstitial elements such as carbon and nitrogen.

Heat treatment in the temperature range 1100–1200°C can have a dramatic influence on the microstructure of a wrought product. Prolonged treatment at high temperatures can lead to equiaxed grains, whereas, cooling at intermediate rates can render the grains acicular, with Widmanstätten morphology. Step quenching, with or without simultaneous mechanical strain can lead to a dual structure, consisting of both coarse and fine austenite grains.

Duplex alloys with high contents of Cr, Mo and W are most susceptible to intermetallic precipitation. Mo and W extend the stability range of intermetallics to higher temperatures. For this reason, higher solution annealing temperatures, i.e. above 1000°C are necessary. In order for the precipitates to re-dissolve, solution annealing temperatures for superduplex grades must be performed at 1050°C and above. For grades such as S32550 and S32750, a few minutes at 1050–1070°C are sufficient, whereas, for tungsten bearing grades such as S32760, 1100°C has been recommended [10]. Li et al. [15], during their investigation of a hyper duplex stainless steel without W additions, solution annealed the samples at 1100°C for 1 h and obtained a microstructure of ferrite and austenite with no secondary phases. Jeon et al. 2012 solution treated a hyper duplex alloy with high W additions at 1090°C for 30 min and found no secondary phases.

Figure 2 shows the SEM micrograph of the as-received S2205 DSS in the wrought form in which austenitic islands are embedded in the ferritic matrix with some undissolved precipitates. Hence, in order to dissolve these harmful precipitates samples were subjected to solution heat treatment by heating it to 1110°C for 60 min. The solution heat treatment was also done to adjust the austenite and ferrite phase proportions which can be seen in the **Figure 3**. If any macro segregations are present in the sample the solution heat treatment will help to eliminate them [16].

5.2. The 475°C embrittlement

Duplex stainless steels (DSSs) have high strength, excellent corrosion resistance and good weldability, and are thus widely used in primary circuit piping of pressurized water nuclear reactors (PWRs). However, after long-term service at temperatures ranging from 280 to 450°C, DSS are subjected to thermal aging embrittlement. The thermal aging embrittlement has caused the degradation of DSS in impact toughness, corrosion properties, and fatigue properties. So far much attention has been paid to the structural integrity assessment and life prediction of aged DSS components subjected to thermal aging embrittlement.

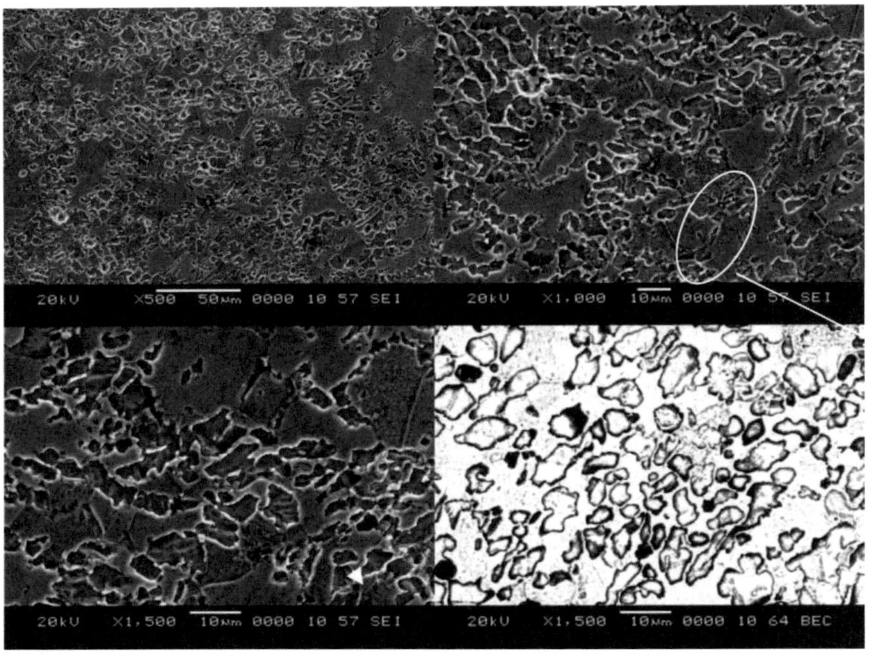

Figure 2. Microstructure of as received S2205 duplex stainless steel sample (Etchant:Beraha's tint etch).

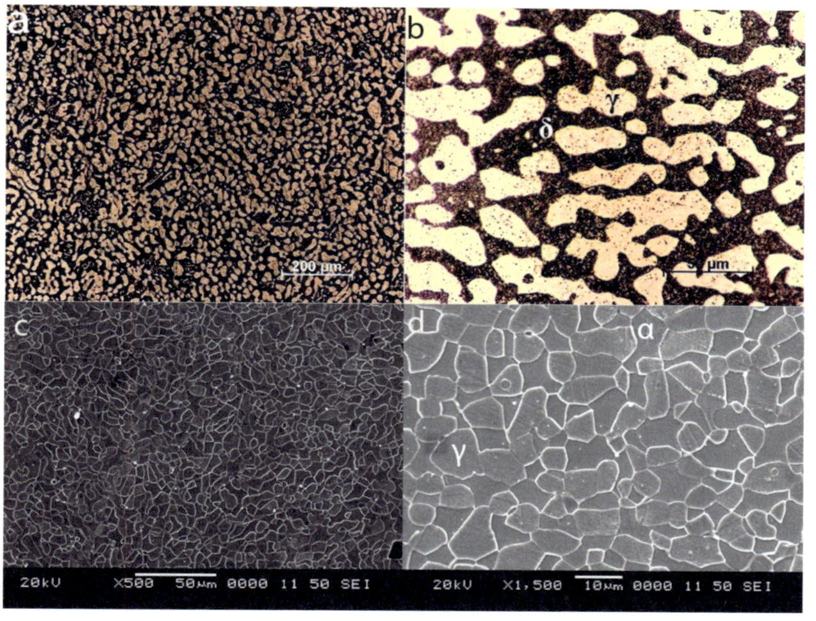

Figure 3. Microstructure of S2205 after solution heat treatment (a) OM image of distribution of austenite grains in the ferrite matrix (magnification: 100×). (b) OM image of phase morphology (magnification: 500×).

Early studies indicate that the thermal aging embrittlement of DSS is due to spinodal decomposition that ferrite phases decompose into coherent Cr-enriched α' and Fe-rich α. It is well known that spinodal decomposition in ferrite of DSS leads to loss in toughness, increase in ferrite hardness and little or no change in tensile properties. In addition, the spinodal decomposition is also accompanied by the chromium nitrides and G-phase precipitation in the ferrite phases.

William [17] was the first to analyze the existence of miscibility gap in iron-chromium phase diagram in the temperature range of 280–500°C because of formation of α' precipitate in the ferritic phase due to the spinodal decomposition as shown in **Figure 4**. According to this phase diagram 475°C embrittlement would occur in the temperatures below 516°C in the composition range of 12–92 at. % chromium in the iron chromium binary alloy system. Blackburn and Nutting [18], Reidrich and Loib [19], completely redissolved the α' in the ferritic steels after 24 h of aging at 550°C. Vintaikin and Loshmanov [20] in his neutron diffraction studies confirmed the clustering and decomposition of ferritic phase in the temperature range of 280–500°C.

Cahn [21] and Hilliard [22] made some critical reviews on the theory of decomposition in metastable ferritic alloys The reviews included the thermodynamical distinction within the miscibility gap in: (a) spinodal decomposition; (b) nucleation and growth of α'. Spinodal decomposition refers to a reaction where two phases of the same crystal lattice type, but different compositions and properties, form due to the existence of a miscibility gap in the alloy system by means of uphill diffusion without nucleation. Thermodynamically this is possible at concentration between the points where the second derivative of the free energy with composition equals zero. This phase separation process occurs at a very fine scale (of the order of

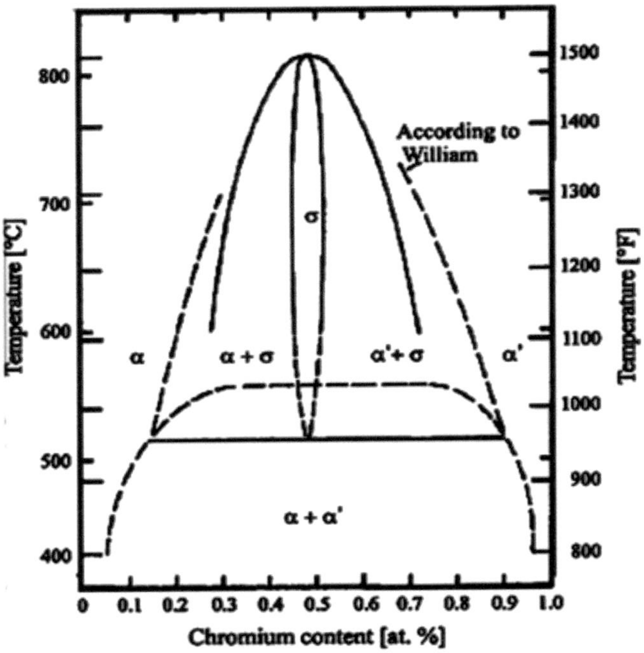

Figure 4. Phase diagram of the iron-chromium binary system.

only a few nanometers) and the presence of the α' phase can only be detected through an atom probe field ion microscope [23, 24]. According to Cahn [20] there should be no change in molar volume with composition in an infinite isotropic solid free from imperfections in order to have spinodal decomposition. So the sustainability of spinodal decomposition is very stringent for a multi component ferritic phase as in DSS, where many alloying elements other than chromium are partitioned to the ferritic phase. Chandra and Schwartz [25] tried to calculate the solubility of chromium in iron in the temperature range of 280–500°C and predicted the boundaries within the miscibility gap for spinodal decomposition, nucleation and growth. The coherent spinodal line in the diagram is the corrected chemical spinodal, taking into account the elastic strain energy due to 0.6% difference in the atomic size of iron and chromium. R.O. Williams. [17] studied the Mossbauer effect of 475°C embrittlement of a series of iron–chromium binary alloys by varying the chromium content and estimated the solubility of chromium in iron to be 12 wt. % at 475°C. They predicted that only alloys with chromium content in excess of 12 wt. % exhibit 475°C embrittlement. Another important observation was that alloys with chromium content 12–30 wt. % decomposed via a nucleation and growth mechanism.

Shamanth et al. found drastic increase in tensile strength and decrease in ductility due to the embrittlement associated with the formation of α and α' precipitates in the ferrite phase for the samples aged at 475°C for 1000 h as depicted in the **Figure 5** where the mottled contrast in the ferrite matrix at nano-scale indicates the presence of Fe-rich α and Cr-rich α' precipitates formed by the spinodal decomposition [5]. It is difficult to distinguish between α and α' precipitate from HR-TEM results even at higher magnifications because both α and α' precipitate had similar lattice parameter as that of Fe [9]. Also in this investigation they found that the α' precipitate associated with the coprecipitation of Cr2N precipitates in the ferrite matrix.

http://dx.doi.org/10.5772/intechopen.80007

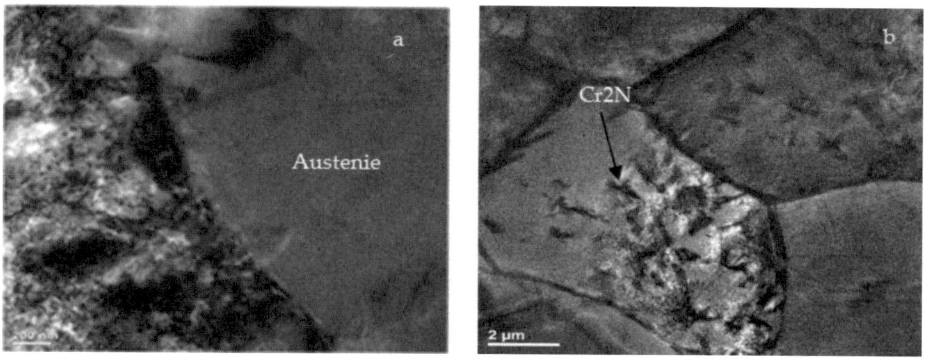

Figure 5. HR-TEM image of 1000 h aged sample showing (a) the mottled contrast ferrite region and precipitate free austenitic phase. (b) the morphology of Cr2N precipitates in ferrite matrix.

5.3. Mechanism of recovery

According to Fe-Cr phase diagram, the spinodal decomposition in ferrite is due to the presence of the miscibility gap in the Fe-Cr system. The spinodal decomposition in the ferrite matrix takes a very long time because of the low atomic mobility at lower temperature (400°C). However the reversion heat treated temperature 550°C is above this (α + α') miscibility gap, the ferrite gets homogenized again or in other words (α + α') will become thermodynamically unstable and dissolve quickly inside the ferrite matrix when they are heat treated at 550°C for shorter duration i.e. maximum of about 60 min. If the time exceeds beyond 60 min there is a possibility for the formation of other intermetallic phases like R-phase, σ-phase or γ-phase. [16]

On subjecting the embrittled specimens to reversion heat treatment at 550°C, it was observed that the strength dropped while the ductility increased, owing to the dissolution of α and α' precipitates into the ferrite matrix as seen in the **Figure 6**. The mechanical properties obtained for Reversion heat treated samples are shown in the **Table 1**. After 30 and 45 min of reversion

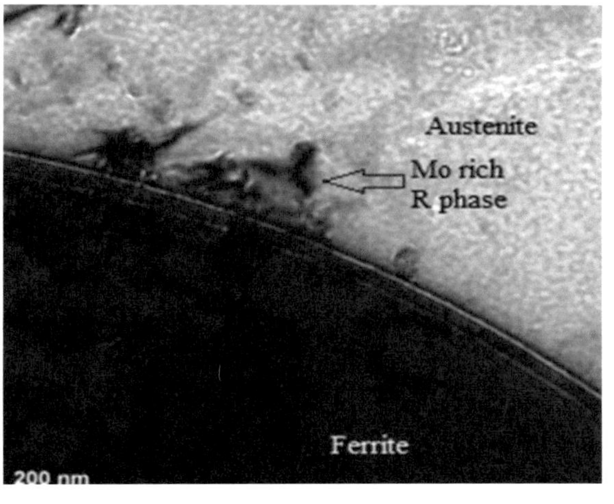

Figure 6. HR-TEM image of 60 min reversion heat treated sample showing clear austenite phase and ferrite phase with R-phase.

Heat treatment	%RE	σ_{ys}(MPa)	%RE	σ_{uts} (MPa)	%RE	Ductility (%)	%RE	Impact value (J)
Solutionized	—	458	—	729	—	42	—	270
Aged-700 h	—	961	—	1261	—	18	—	6
RHT-15 min	20	730	46	935	61.3	31	45.8	140
RHT-30 min	28.9	609	70	842	78.8	38	75	190
RHT-60 min	94.3	528	86	780	90.4	39	80	250
RHT-90 min	36.6	610	69.7	785	89.4	34	96	200
RHT-120 min	26.8	632	65.4	854	76.5	34	96	190
Reaged-25 h	—	502	—	802	—	25	—	90
Reaged-50 h	—	548	—	825	—	24	—	80
Reaged-75 h	—	609	—	892	—	24	—	78
Reaged-80 h	—	861	—	1032	—	19	—	10
Reaged-90 h	—	982	—	1125	—	19	—	6
Reaged-100 h	—	993	—	1208	—	18	—	6

Table 1. Mechanical properties of all heat treatments conducted.

treatment, the decrease in strength was marginal, whereas, the increase in ductility was significant. The material exhibited a further drop in strength and increase in ductility for 60 min of reversion, after which, both the strength and ductility remained more or less constant. The reversion treatment, however, was unable to completely restore the properties of the material back to those corresponding to the un-aged or solution annealed condition. The yield strength was more sensitive to reversion heat treatment, varying much more than the tensile strength with respect to duration of reversion heat treatments. It is because of the further embrittlement that occurred simultaneously with the dissolution of α and α' precipitates, during reversion heat treatment and, as a result, the mechanical properties were not fully recovered. The temperature range of the reversion heat treatment is favorable for the formation of a Mo-rich precipitate called R-phase, which is known to be an embrittling phase [11], and in fact, previous investigators [12], had found unidentified Mo-rich precipitates in super-duplex stainless steels reversion treated for 30 min at the same temperature as in the current investigation. Another possible reason for this behavior could be residual inhomogeneities in the ferrite grains as a result of the Cr and Mo atoms that were part of the precipitates might have been redistributed. This in turn may have resulted in the formation of dislocation substructures in the ferrite grains during deformation. From the results it was also very clear that after 75 min reversion treatment the strength was again increasing and the ductility was decreasing because of the embrittlement caused by the R-phase and secondary austenite which were quantitatively seen in **Figure 6** (TEM micrograph for 120 min reversion heat treated sample). The measured values of ultimate tensile strength and % elongation for re-aged conditions are given in **Table 1**.

6. Conclusion

From the studies we can conclude that the DSS are susceptible for various intermetallic phases when subjected to temperatures of around 300–500 °C. The intermetallic phases formed are Fe-rich α and Cr-rich α' precipitate because of the spinodal decomposition and also some amount of Cr_2N precipitates were seen in the aged ferritic matrix. From the reversion heat treatment technique, the alpha and alpha-prime precipitates was completely dissolved. However, when the embrittled samples are subjected to optimum reversion heat treatment temperature and time, the alpha and alpha-prime precipitates can be completely dissolved.

Author details

V. Shamanth[1], K. S. Ravishankar[2] and K. Hemanth[1]*

*Address all correspondence to: ckh4545@gmail.com

1 School of Mechanical Engineering, Reva University, Bangalore, India

2 National Institute of Technology Karnataka, India

References

[1] Lippold JC, Kotecki DJ. Welding Metallurgy and Weldability of Stainless Steels. New Jersey: John Wiley and Sons, Inc; 2005

[2] Ki Leuk Lai J, Shek HS, Ho Lo K, editors. Stainless Steels: An Introduction and Their Recent Developments. Sharjah: Bentham Science Publishers; 2012

[3] Davis JR. Stainless Steels. ASM International; 1994

[4] Kazior J, Nykiel M, Pieczonka T, Puscas TM, Molinari A. Activated sintering of P/M duplex stainless steel powders. Journal of Materials Processing Technology. 2004;**157-158**:712-717

[5] Alvarez-Armas I, Degallaix-Moreuil S, editors. Duplex Stainless Steels. New Jersey: John Wiley and Sons, Inc.; 2009

[6] Chung HM. Aging and life prediction of cast duplex stainless steel components. International Journal of Pressure Vessels and Piping. 1992;**50**(1):179-213

[7] Chen TH, Weng KL, Yang JR. The effect of high-temperature exposure on the microstructural stability and toughness property in a 2205 duplex stainless steel. Materials Sc ience and Engineering: A. 2002;**338**:259-270

[8] Konosu S. Effect of reversion heat treatments on the mechanical properties of a 13% Cr steel subjected to 475°C embrittlement. Scripta Metallurgica et Materialia. 1992;**26**(10):1631-1636

[9] Jang H, Hong S, Jang C, Lee JG. The effects of reversion heat treatment on the recovery of thermal aging embrittlement of CF8M cast stainless steels. Materials and Design. 2014; **56**:517-521

[10] Gunn RN. Duplex Stainless Steels: Microstructure, Properties and Applications. Woodhead Publishing; 1997

[11] Redjaimia A, Metauer G, Gantois M. Decomposition of Delta ferrite in an Fe--22 Cr--5 Ni--3 Mo--0.03 C duplex stainless steel. A morphological and structural study. Duplex Stainless Steels'91. 1991;**1**:119-126

[12] Nilsson JO. Overview: Super duplex stainless steels. Materials Science and Technology. 1992;**8**(8):686-700

[13] Hwang TH, Kim JH, Kim KH, Moon WJ, Kang CY. Effect of R-phase on impact toughness of 25Cr-7Ni-4Mo super duplex stainless steel. Metals and Materials International. 2014;**20**(1):13-17

[14] Davison. Practical guide to using duplex stainless steels. Materials Performance. 1990; **29**(1):57-62

[15] Li SL, Zhang HL, Wang YL, Li SX, Zheng K, Xue F, Wang XT. Annealing induced recovery of long-term thermal aging embrittlement in a duplex stainless steel. Materials Science and Engineering: A. 2013;**564**:85-91

[16] Shamanth V, Ravishankar KS. Dissolution of alpha-prime precipitates in thermally embrittled S2205-duplex steels during reversion-heat treatment. Results in Physics. 2015; **5**:297-303

[17] Williams RO. Further studies of the Iron-chromium system. Trans. TMS-AIME. 1958;**212**: 497-502

[18] Blackburn MJ, Nutting J. Metallography of Fe-21%Cr alloy subjected to 457 °C embrittlement. Journal of the Iron and Steel Institute. 1964;**202**:610-613

[19] Reidrich G, Loib F. Archiv fur das Eisenhuttenwesen. 1941;**15**:175-182

[20] Vintaikin EZ, Loshmanov AA. Brittleness of 475 c. Iron-chromium alloys. Fiz Metallov Metalloved. 1966;**22**(3):473-476

[21] Cahn JW. On spinodal decomposition. Acta Metallurgica. 1961;**9**(9):795-801

[22] Hilliard JE. In: Aaronson HI, editor. Phase Transformations. Ohio: American Society for Metals, Metals Park; 1970

[23] Johansson J, Odén M. Load sharing between austenite and ferrite in a duplex stainless steel during cyclic loading. Metallurgical and Materials Transactions A. 2000;**31**(6):1557-1570

[24] Grobner PJ. The 885 F (475 C) embrittlement of ferritic stainless steels. Metallurgical Transactions. 1973;**4**(1):251-260

[25] Chandra D, Schwartz LH. Mössbauer effect study of the 475‡ C decomposition of Fe-Cr. Metallurgical Transactions. 1971;**2**(2):511-519

Chapter 6

Abrasion-Corrosion of Ferritic Stainless Steel

Wilian S. Labiapari, Miguel A. N. Ardila,
Henara L. Costa and José Daniel B. de Mello

Additional information is available at the end of the chapter

http://dx.doi.org/10.5772/intechopen.81913

Abstract

Several studies have measured abrasion-corrosion for biomaterials, alloys, and stainless steel. Despite the considerable effort to understand the synergy between abrasion-corrosion resistance of stainless steel, they have mainly focused on more traditional materials, such as AISI 304 and AISI 316 stainless steel, and, more recently, on AISI 2205 duplex stainless steel. Little progress has been made to understand this phenomenon for cost-effective ferritic stainless steel. In this chapter, we first show the great potential of the use of ferritic stainless steel in the sugar cane biofuel industry. The influence of their crystallographic texture on the corrosion resistance of 16% Cr ferritic stainless steel (both niobium-stabilized and non-stabilized) is presented and discussed. We also analyses the microabrasion-corrosion performance of ferritic stainless steel with different chemical compositions (11%wt Cr with and without Ti stabilization; 16%wt Cr with and without Nb stabilization) and, for comparative purposes, austenitic stainless steel (18%wt Cr-8%wt Ni) and carbon steel (0.2%wt C). For all materials tested, microabrasion wear coefficients were higher (4x) than those measured under abrasion-corrosion conditions. Friction coefficients could also be measured by a 3D load cell strategically positioned in the specially developed microabrasion-corrosion device, showing a strong reduction (2x) in friction coefficient under abrasion-corrosion conditions when compared with solely abrasion conditions.

Keywords: ferritic stainless steel, abrasion-corrosion, synergy, force measurements, friction coefficient, potentiodynamic polarization, crystallographic texture, Ti and Nb stabilization, biofuel industry

1. Introduction

In various engineering applications, mechanical components are simultaneously subjected to a combination of mechanical wear and corrosion [1, 2]. Due to their high corrosion resistance, stainless steel could be an interesting candidate, despite its relatively low mechanical resistance when compared to other hard wear-resistant materials. For example, in conditions of moderate corrosion in combination with mechanical wear, such as in the initial stages of sugar cane plants for ethanol production, a previous work proved an exceptional performance of inexpensive ferritic stainless steel at low cost-benefit ratios [3].

The corrosion resistance of stainless steel is attributed to the formation of a protective passivated layer, and they are generally regarded as materials that are easily repassivated. The dynamics involving the removal of the passive layer by mechanical action and repassivation plays an important role in the abrasion-corrosion resistance of stainless steel. The complex tribochemical mechanisms of stainless steel depend on the microstructure and chemical composition of the material surface, the solution pH, the abrasive (size, type and concentration) and the imposed electrochemical conditions [4].

Most studies about abrasion-corrosion resistance of stainless steel are mainly focused on more traditional materials, such as AISI 304 and AISI 316 stainless steel, and, more recently, on AISI 2205 duplex stainless steel. The investigation of cost-effective ferritic stainless steel for those applications is often neglected. Despite this, they have found an important application niche in the biofuel industry [3], which certainly involves abrasion and corrosion. Within ferritic stainless steel, it is relevant to investigate the effect of Cr content on tribocorrosion, as well as the effect of stabilization. The cheapest stainless steel is 11Cr (DIN 14003), and it has vast application in the sugar cane biofuel industry. 11CrTi (ASTM S40910) is also a low-cost ferritic stainless steel stabilized with Ti, largely used in automobile exhaust systems. 16Cr (ASTM S43000) steel is a slightly more expensive ferritic stainless steel, mainly used in the cutlery industry, but it is still cheaper than austenitic steel. 16CrNb (ASTM S43000) is the same 16Cr stainless steel stabilized with Nb used in cutlery and stamping. On the other hand, austenitic stainless steel, such as 18Cr8Ni (AISI 304), presents higher cost, but is extremely versatile in its use, with high corrosion resistance, good formability and weldability.

The abrasion and corrosion phenomena have become of great importance in sectors where the contact between two surfaces in relative motion and the chemically reactive environment are the main failure factors. Thus, several methodologies and tests have been developed to study the mechanisms that originate the phenomena of abrasion and corrosion separately; but in the industry, the phenomena of abrasion and corrosion happen simultaneously. Some examples of such occurrences are pumps, process valves, oil-gas pipelines in the marine industry and metal surgical implants replacing bone parts of the human body [1]. With advances in the understanding of abrasion and corrosion, independently treated, the interest and necessity of studying the microabrasion interactions in aqueous conditions arose, especially when the environment is corrosive [1, 2].

In systems where corrosion and mechanical wear co-occur, phenomena such as plastic deformation not only influence mechanical failure due to wear, but also lead to the removal of passivation layers present on the metal surface. Exposed metal surfaces can be highly reactive to the environment, which can accelerate corrosion [5].

The combined effect of abrasion and corrosion, the so-called synergistic effect, has been widely debated in the literature. Corrosion has been shown to accentuate abrasion [2, 4, 6–9], while for others abrasion is attenuated by corrosion [7, 9]. However, even though considerable efforts have been made to understand the synergy between abrasion and corrosion, little progress has been made in quantifying this phenomenon for stainless steel [6].

2. Ferritic stainless steel as an antiwear material for the biofuel industry

Society and governments have been highly interested in a large-scale production of alternative forms of energy, such as biofuels, which reduce greenhouse gas emissions and improve energy security when compared to their fossil counterparts. As a successful example, the vast majority of automotive cars in Brazil (86%) are manufactured with engines that can use any amount of gasoline or ethanol, a technique called flex-fuel [10]. One of the most productive sources of ethanol is sugar cane. The energy matrix based on biofuels in Brazil started in the 1980s [11], when the production of ethanol was 2500 l/hectare. Nowadays, the production is 6500 l/hectare and it is expected to reach 13,000 l/hectare in 2020. With increased production and efficiency, the period available for the annual plant maintenance has been drastically reduced. One of the main reasons for the annual maintenance is premature wear of devices used to wash, cut and crush sugar cane. In general, the main material used in sugar cane plants is structural steel with low carbon content. The main rational for this choice is the material's low cost, although its good weldability is also an important factor. It is expected that abrasive wear will occur, mainly due to the presence of cane husk, soil and sand present in the cane, but that the severity of abrasive wear will be moderate. Therefore, the wear life of the components should allow them to last for the whole sugar cane season, and their majority would be replaced for the next season. Corrosion in the components used to receive, wash, cut, shred and mill sugar cane could be relevant both during season and during off-season. During season, corrosion due to humidity and water (cane washing, rain, dew, added water, etc.) could occur concomitantly with abrasive wear due to cane husk, soil and sand particles transported with the cane and the synergy between abrasion and corrosion should be investigated. During off-season, corrosion due to humidity and water (rain, dew) probably does not passivate carbon steel and leads to irregular surfaces with corrosion products, which will then be subsequently removed by abrasive wear in the next season.

A recent paper proposed and evaluated an alternative solution for the biofuel industry [3]. It investigates the use of low-cost ferritic stainless steel in real tribocorrosive systems subjected to corrosion due to water and moderate abrasive wear simultaneously. The first step was a pioneer introduction of stainless steel in some components used in the initial stages of sugar cane processing, motivated by corrosion that was easily observed in the components during off-season.

A visual comparison between carbon steel (A36) and ferritic stainless steel (P410) is shown in **Figure 1**, where cane transportation conveyors made of the two materials are compared. Since the samples were collected during season, the movement of the cane on the surfaces helps to clean corrosive products and, therefore, both surfaces appear smooth and clean. However, a

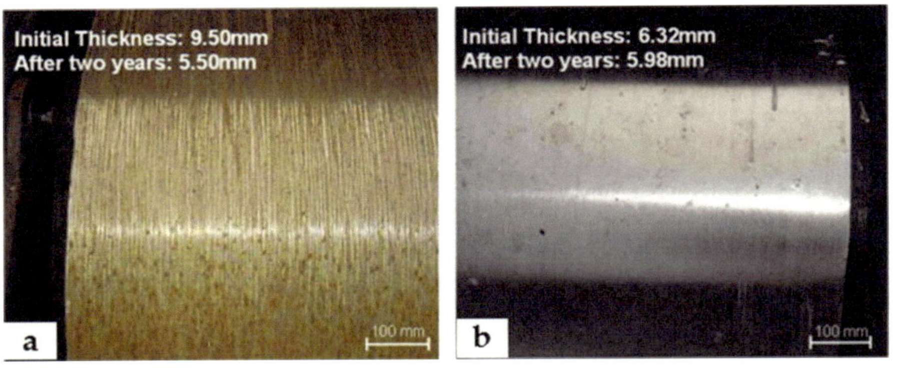

Figure 1. Worn surfaces in cane transport conveyors during field tests after use during two seasons: (a) A36 and (b) P410D [3].

yellow coloration can be observed for carbon steel because the samples are presented after 24 h of maintenance, showing that even during a shortstop, corrosion products begin to accumulate for carbon steel [12].

In addition, the wear measured by the reduction in thicknesses of the carbon steel and ferritic stainless steel sheets installed in similar components is summarized in **Figure 2**. The dotted lines refer to the thickness reduction of a carbon steel sheet and the complete lines refer to ferritic stainless steel sheet. The superior performance of stainless steel is clearly evidenced, where the thickness reduction was minimal. The locations 1, 2, 3 and 4 indicated in this figure relate to the four positions along the sheet [3].

The change to ferritic stainless steel also influenced the quality of the sugar produced. The introduction of ferritic stainless steel in sugar cane industries started only for a few components in 2003 and gradually increased until 2007, so that in 2007 the vast majority of the

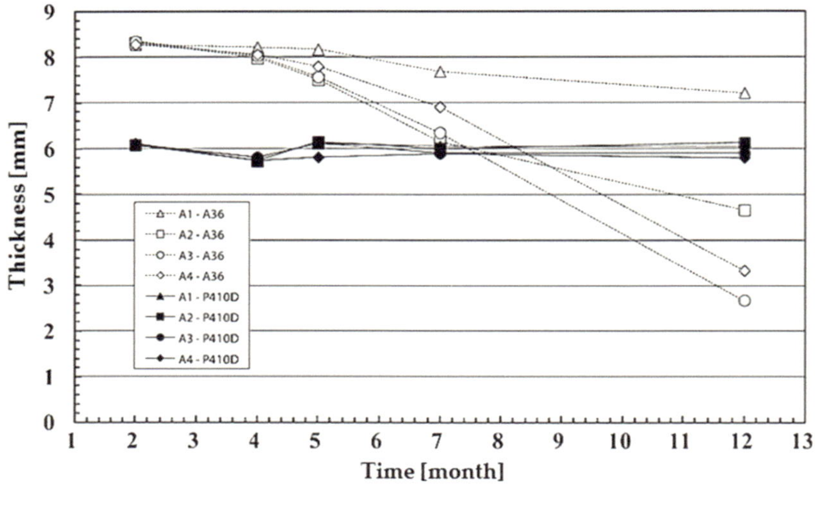

Figure 2. Thickness reduction of carbon steel and stainless steel plates during 10 months of use in field tests [3].

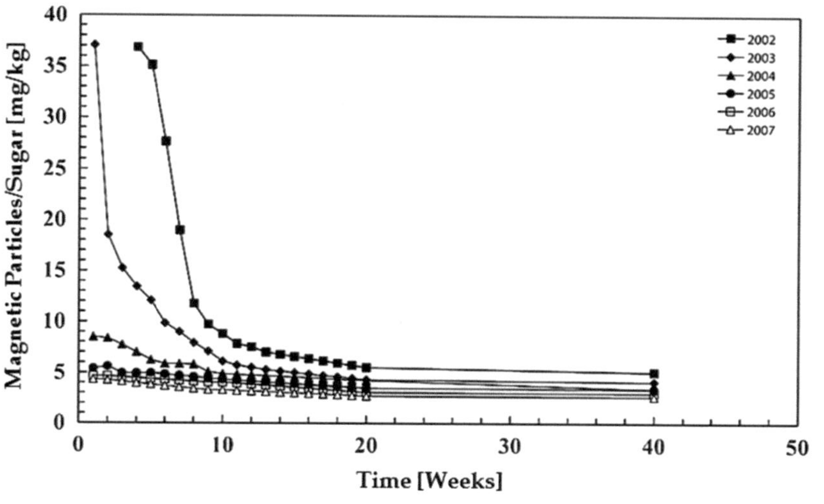

Figure 3. Evolution of the amount of magnetic particles in sugar produced between 2002 and 2007 [3].

components used to receive sugar cane and move it to the processing line were replaced by steel stainless. **Figure 3** shows the comparison of the evolution of the weight of magnetic particles in the final product due to wear debris between the years of 2002 and 2007. This figure clearly shows that before 2003, a few weeks of use were necessary to significantly reduce the number of debris in the final product, while in 2007, good quality sugar (without metallic wear debris) could be produced in the first week of use [3].

Carbon steel parts (which have been gradually reduced from 2003) can suffer severe corrosion during off-season or maintenance stops because carbon steel does not passivate [12, 13]. Corroded areas will likely become preferred locations for abrasive wear and the synergy between abrasion and corrosion will lead to severe material removal. When carbon steel was completely replaced by stainless steel (2007), the corrosion products accumulated during the off-season could be removed in the first week of use, for the passivation protects the surface against corrosion caused by rain, dew, etc. The authors [3] point out that when carbon steel is used, even after reduction to a much smaller level, the amount of metal debris in the final sugar is still significantly higher than when stainless steel is used, which shows a synergy between abrasion and corrosion also during season, when the sliding of the surfaces continuously removes possible products of corrosion [3].

3. Influence of crystallographic texture and niobium stabilization on the corrosion resistance of ferritic stainless steel

Previous studies reported that after hot and cold rolling processes, stainless steel showed preferential crystallographic textures [14–20]. In particular, Raabe and Lücke [17] showed that ferritic steel, both non-stabilized and stabilized with Nb or Ti, had a texture gradient throughout the thickness of the samples. It was observed that the magnitude of the Goss shear texture varied with the position throughout the thickness, specially prevailing at 20% depth from the sheet surface. The Goss texture occurs as a recrystallization texture for FCC materials. For ferritic steel AISI 430 with different Nb content, new crystallographic texture components appear and these are attributed to the formation of coarse grains when niobium content varies [21]. Crystallographic texture analyses were performed in a conventional "Electron Backscatter Diffraction" (EBSD) system attached to a SEM. **Figure 4** exemplifies, using EBSD analysis, the crystallographic texture distribution throughout the thickness of AISI 430 rolled ferritic stainless steel with (16CrNb) and without Nb (16Cr) stabilization (see **Table 1** for chemical composition). The ferritic stainless steel samples' thickness was 4 mm. **Figure 4a** shows the scheme used for extraction of steel samples from the regions defined as surface and centre.

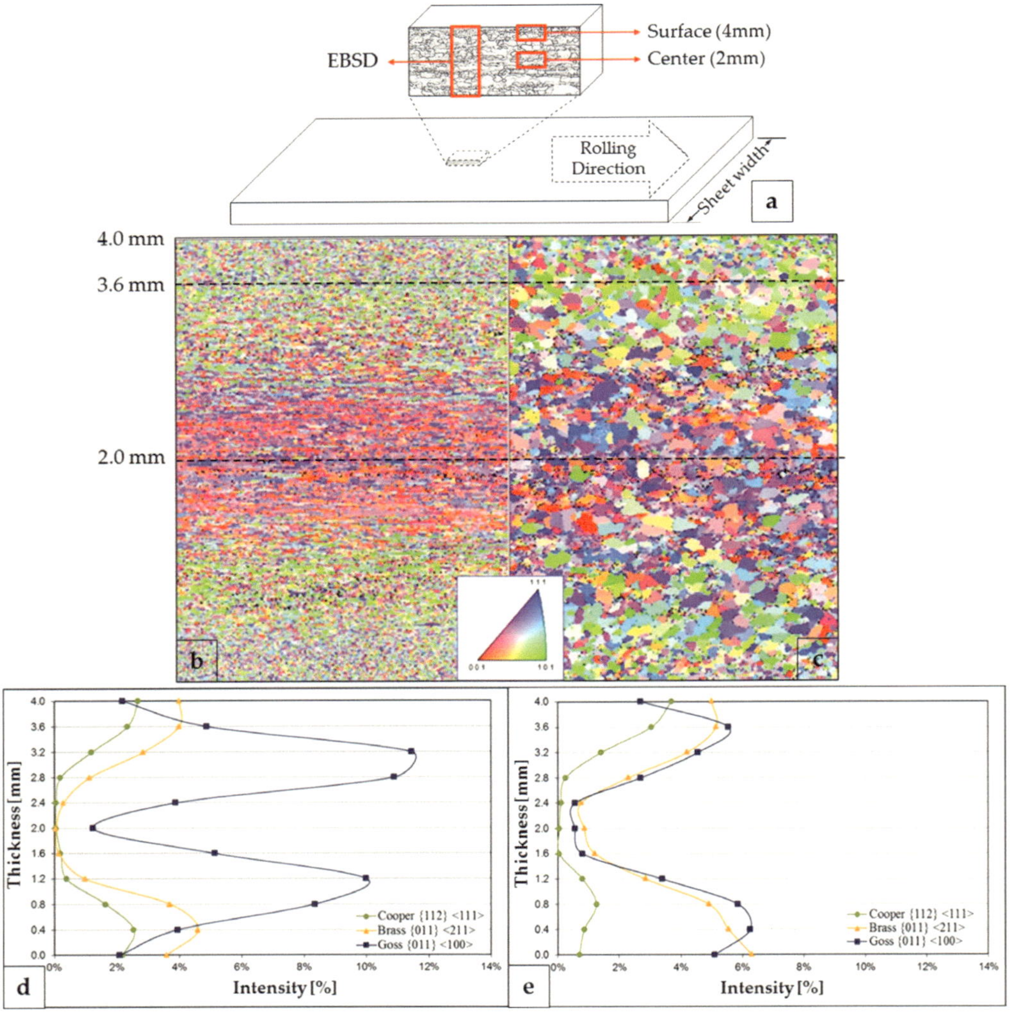

Figure 4. EBSD analysis of a 16Cr ferritic stainless steel with and without Nb stabilization: (a) Scheme of sample extraction; Inverse pole figure: (b) 16Cr, (c) 16CrNb; Texture intensity through the thickness: (d) 16 Cr, (e) 16 CrNb.

Element (%wt)	11Cr	11CrTi	16Cr	16CrNb	17CrTiNb	18Cr8Ni	A36
C	0.011	0.009	0.049	0.025	0.011	0.055	0.138
Mn	0.61	0.13	0.32	0.21	0.16	1.15	1.06
Si	0.49	0.52	0.36	0.46	0.36	0.42	0.01
P	0.0247	0.0182	0.0395	0.0358	0.0219	0.0251	0.0154
S	0.0002	0.0005	0.0015	0.0013	0.0024	0.0009	0.0075
Cr	11.23	11.29	16.10	16.19	17.62	18.28	0.01
Ni	0.31	0.12	0.27	0.19	0.17	8.01	0.01
Mo	0.021	0.005	0.020	0.035	1.75	0.063	0.003
Al	0.002	0.003	0.002	0.002	0.0055	0.003	0.033
Nb	0.006	0.002	0.014	0.416	0.183	0.005	0.001
Ti	0.003	0.144	0.003	0.004	0.177	0.001	0.001
N	0.0145	0.0087	0.0528	0.0202	0.0112	0.0421	0.0026

Table 1. Chemical composition of the specimens.

The inverse pole figures obtained for the EBSD texture analyses are given in **Figure 4b** and **c**. It can be seen that 16Cr samples show a larger texture gradient throughout their thickness when compared to 16CrNb samples. On the surface, the preferential orientations are <101> and <111>, whereas in the centre, they are <001> and <111>. It was also inferred that the 16CrNb steel has larger grains than the 16Cr steel. However, Ardila et al. [21] argued that the difference in grain size between 16Cr and 16CrNb steel had no influence on pitting potential.

For a quantitative description of the crystalline orientations, a crystalline orientation distribution function (CODF) was used. Only shear textures were considered, see **Figure 4d** and **e**. It can be seen that the 16Cr and 16CrNb steel have similar textures at positions that corresponded to different thicknesses (4, 3.6 and 2 mm). It is also observed that in the centre of the plate (position corresponding to 2 mm), there are no relevant shear textures—all have intensities lower than 1%. The principal shear texture at the position corresponding to 3.6 mm was Goss 011 <100> with intensity 4.9% for the 16Cr and 5.5% for the 16CrNb steel. It can also be observed that the shear texture brass 011<211> is more intense closer to the surface (position corresponding to 4 mm). This is due to the recrystallization taking place preferentially at the metal surface and can be explained by the plastic deformation generated during this phenomenon. Deformation is influenced by the presence of alloying elements such as niobium [17, 20, 22], and explains the slightly high intensity of the brass texture for steel 16CrNb (5%) compared to the steel 16Cr (4%). The copper shear texture 112<111>/shear texture Goss 011<111> was present with similar intensity at the positions corresponding to the surface (4 mm) and 3.6 mm, with approximately 2.50% for the 16Cr steel and 3.35% for the 16CrNb steel. The Goss texture also prevails in the 16Cr steel, but its highest intensity occurs at the thickness of 3.2 mm (65% from the centre to the surface), although at 3.6 mm the values are also high. The difference in intensity throughout the thickness may be caused by the stabilization with niobium [17, 20, 22], and also by the grain size [22]. Summarizing, the shear texture Goss 011 <100> is the most important texture variation between the surface (4 mm) and the bulk of the plate. For both materials, the fibre texture was little affected by the position [19]. As a consequence, we may suppose that variation in properties is governed by the presence of Goss texture with greater intensity at 3.6 mm.

To verify the influence of the crystallographic texture on corrosion resistance, the pitting potentials were measured using samples taken from the same plate for each steel at the positions corresponding to 4, 3.6 and 2 mm, as indicated in **Figure 4a**. Typical results of the anodic potentiodynamic polarization curves (obtained according to ASTM standard G59-97 [23] in a 3.56% NaCl electrolyte) are presented in **Figure 5**.

The 16CrNb steel presented higher pitting resistance than 16Cr steel. The carbon content of 16Cr steel is higher than that of 16CrNb steel (**Table 1**). In this way, 16Cr steel is more susceptible to the formation of Cr carbides, mainly at the grain boundaries. In addition, in the steel stabilized with Nb, carbon will preferentially form Nb carbides instead of Cr carbides, thus decreasing Cr depletion in the matrix [24]. Higher Cr-free content in stainless steel benefits the formation of more stable passive films on the steel surface, which prevents the penetration of chloride and sulfate ions. Consequently, Cr enhances pitting corrosion resistance and uniform corrosion resistance [25]. Moreover, Nb contributes to corrosion resistance when it is added to the alloy [26–29], the addition of this element in ferritic stainless steel changes the characteristics of the surface film of oxide semiconductor caused by Nb^{5+} incorporation into the passive layer. This fact shifts the current to lower values, and results in increased pitting corrosion resistance [26, 29].

Additionally, the pitting potential increases at the position corresponding to 3.6 mm and decreases at the center of the sample (2 mm), that is, the corrosion resistance is lower at the centre. Ardila et al. [21] reported that the orientation of the fibre textures <110> (α, ζ and ε fibers) was predominant for both steel specimens throughout the thickness, whereas the shear texture (Goss) with orientation <100> appeared predominantly at 3.6 mm (**Figure 4**). It is reasonable to suppose that the presence of the orientation <100> in the shear texture is at the origin of the improvement of pitting resistance in ferritic steel, which has higher atomic density intrinsic of the cubic crystalline systems [30–32]. Previous studies [32, 33] showed that

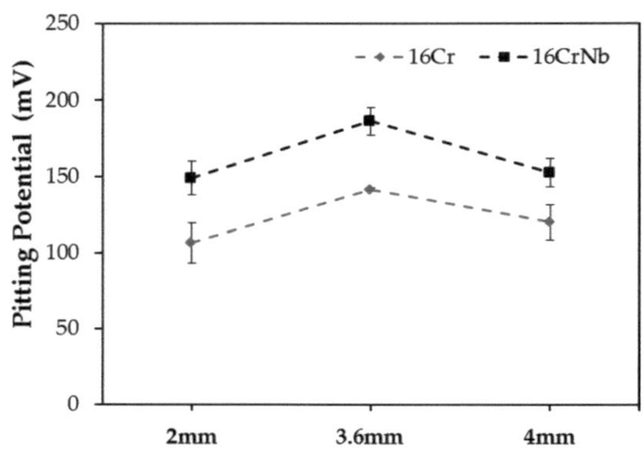

Figure 5. Pitting potential during standard corrosion tests as a function of thickness in 16Cr ferritic stainless steel, with and without Nb stabilization [21].

the lower susceptibility to pitting is related to increasing atomic density planes of the FCC system, with this susceptibility decreasing in the following order: 110 > 100 > 111. The crystallographic planes with a high number of nearest neighbor atoms require higher total energy for the breaking of the bonds and the subsequent dissolution of atoms [34].

Figure 6 shows typical surfaces after anodic potentiodynamic polarization tests in 1 N H_2SO_4. It is clear that for the 16CrNb steel samples, the severity of corrosion allowed a neat visualization of

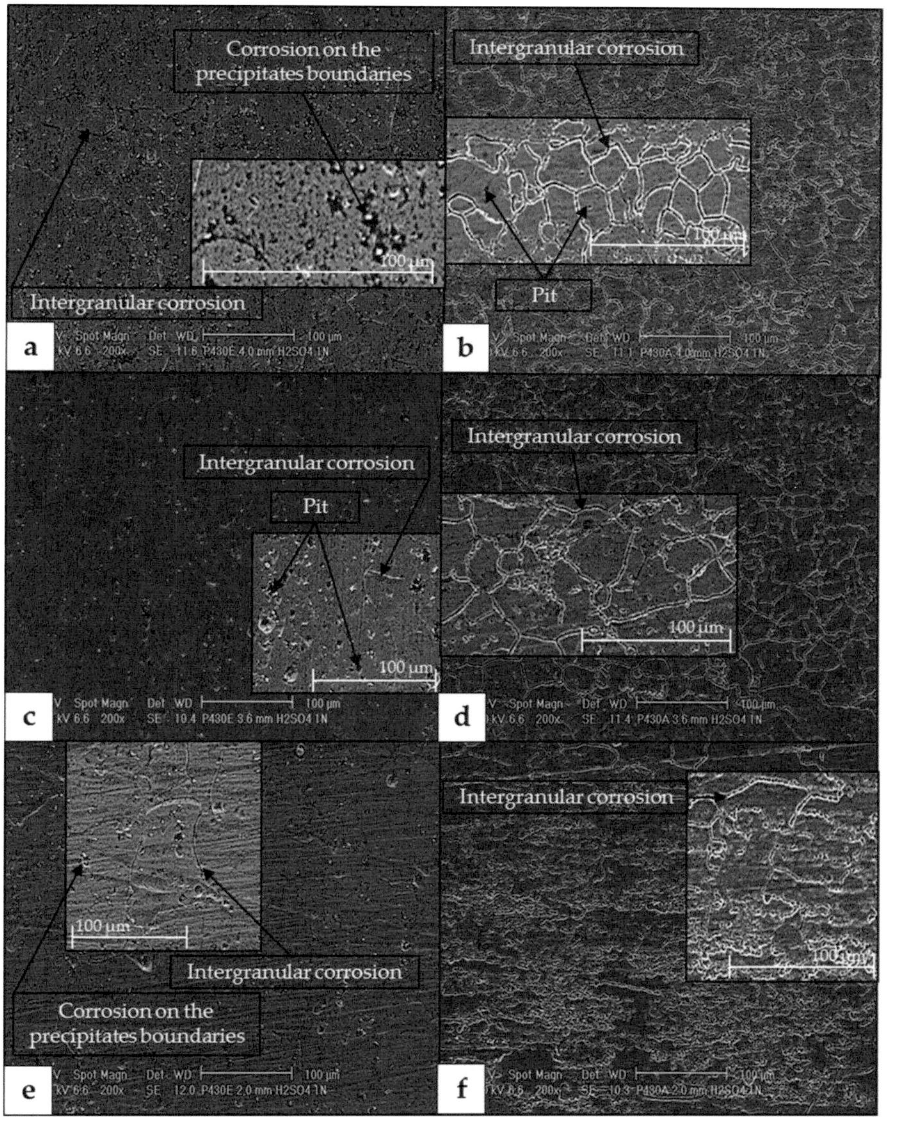

Figure 6. SEM images of the ferritic stainless steel samples tested throughout the thickness in 1 N H_2SO_4. (a) 16CrNb 4 mm; (b) 16Cr 4 mm; (c) 16CrNb 3.6 mm; (d) 16Cr 3.6 mm; (e) 16CrNb 2 mm; (f) 16Cr 2 mm [21].

niobium carbides, whereas for the 16Cr steel samples, intergranular corrosion was more prevalent, which agreed well with previous results regarding corrosion phenomena in 16CrNb and 16Cr steel. Niobium carbide precipitates were observed throughout the surface of the 16CrNb steel samples as white spots. At 4 and 2 mm, weak granular corrosion was identified (**Figure 6a** and **e**), as well as more substantial corrosion at the boundaries of the carbides, a phenomenon that was not detected at 3.6 mm (**Figure 6c**). At this position, in addition to a small amount of intergranular corrosion, a slight presence of pits was observed, but it cannot be concluded that the corrosion is significant there. For 16Cr steel, intergranular corrosion was observed throughout the thickness, although at the surface (position corresponding to 4 mm), some pits could be observed inside the grains (**Figure 6b**). Compared to the centre of the samples, that is, at 2 mm (**Figure 6f**), a significant amount of intergranular corrosion was observed, as well as a generalized degradation of the surface that can be considered generalized corrosion.

4. Microabrasion-corrosion of ferritic stainless steel

With the basic understanding of the influence of systemic factors on the electrochemical corrosion of ferritic stainless steel, the understanding of the influence of abrasion on corrosion and vice versa can be discussed. For this, an "in situ" technique that simultaneously assesses abrasion and corrosion was developed, and with that, a test rig that joins the techniques of microabrasion and electrochemical corrosion was developed [2, 35–40]. The principle is to install an electrochemical cell in a fixed-sphere microabrasive test rig; the cell is connected to a potentiostat for the potential difference application, as exemplified in **Figure 7**. One of the latest hybrid abrasion-corrosion test rigs combined an electrochemical cell and a potentiostat with a fixed ball microabrasion tester [38], and in this way, the contact forces (normal and frictional forces) were monitored in real time. A schematic view of this test rig is shown in **Figure 8**. The equipment can be divided into four parts:

- **Abrasive slurry preparation:** It consists of an electromagnetic mixer that guarantees the homogeneity of the solution and a peristaltic pump to sustain a controlled flow of the slurry during the test. The slurry is a mixture of the abrasive particles and the electrolytic medium.
- **Image acquisition and processing system**: A digital camera ensures the measurement of the wear scar which is analysed by an image processing software.
- **Motion and loading measurements**: The load can be applied using dead weight or an electromagnetic actuator; therefore, it is possible to run a test with a fixed normal load or a variable controlled load. This latter characteristic is quite unique because it allows the comparison of wear behavior under constant load and constant pressure. The force is transmitted from its point of application to the sample by a lever arm system. A three-dimensional load cell (measuring three forces and three moments simultaneously) was installed after the lever arm. In order to protect the load cell against corrosion, it was positioned outside the electrochemical cell and the load was transmitted, using a polymeric rod, through a flexible membrane acting as a cell wall. The flexible membrane poses negligible mechanical

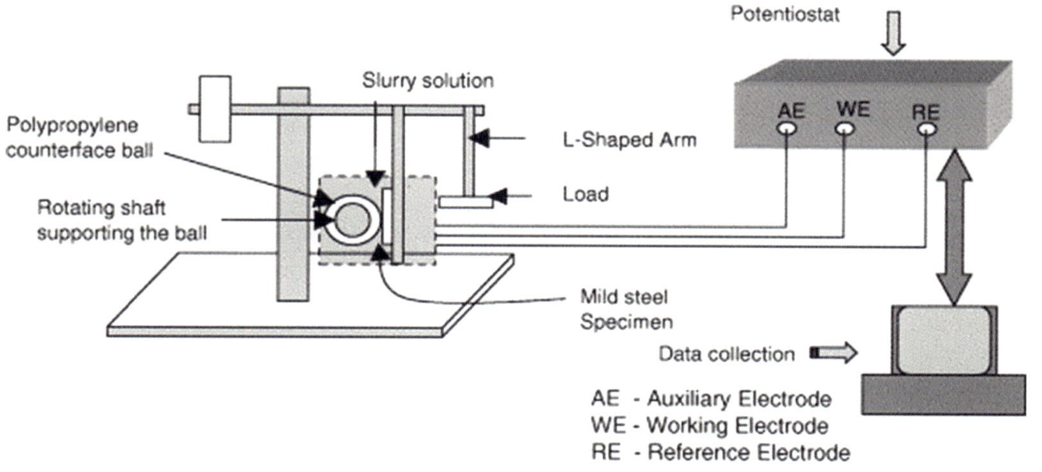

Figure 7. Schematic drawing of the adaptation of the microabrasion test for abrasion-corrosion tests [2].

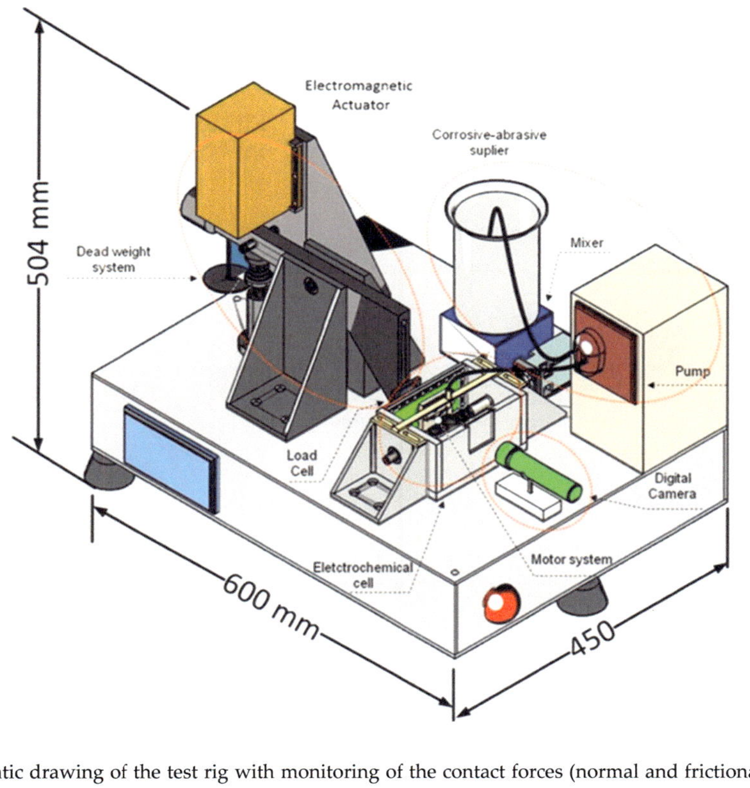

Figure 8. Schematic drawing of the test rig with monitoring of the contact forces (normal and frictional forces) in real time [38].

interference on the load transmission while assuring a watertight and insulated electrochemical cell. The rotary speed of the sphere is assured by a DC motor controlled using a closed loop methodology, where the velocity feedback signal is provided by a quadrature encoder. The possibility of monitoring the efforts throughout the test is a differential in relation to other test rigs reported in the literature.

- **Electrochemical cell**: The electrodes from a Biologic potentiometer, model SP-150, had been positioned within the electrochemical cell, which is made of methyl methacrylate, at a constant level of electrolyte, which assures electrical contact between all electrodes.

The development of these test rigs allows the comparison of the polarization curves in pure corrosion and abrasion-corrosion tests. **Figure 9** exemplifies this comparison for 16CrNb ferritic stainless steel.

The corrosion test conditions were similar to the conditions described in the tests reported in **Figure 5**, that is, they have been aerated and partially submerged and used a saline bridge (turbulent). For the same electrochemical parameters, an increase in the passive current density was observed in the abrasion-corrosion tests when compared with the pure corrosion tests.

This increase in passivation current density is attributed to the dynamics involved in passive layer removal and repassivation [2, 4, 41]. The slight increase in current density in the "passivation region" reflects a less effective repassivation [36]. Evidence of this constant removal and generation of the passive layer is the fluctuation (noise) observed in the current density in the abrasive-corrosive tests when compared with corrosive tests. The fluctuation in the current is due to spontaneous repassivation within the wear scar and subsequent depassivation due to the action of abrasive particles [4]. During the sliding wear of passive metals, the abrupt variation (noise) of the anodic current corresponds to the rate of electrochemical removal of the metal in the wear mark [42].

This chapter analyses the microabrasion-corrosion performance of ferritic stainless steel with different chemical compositions, in particular Cr content:

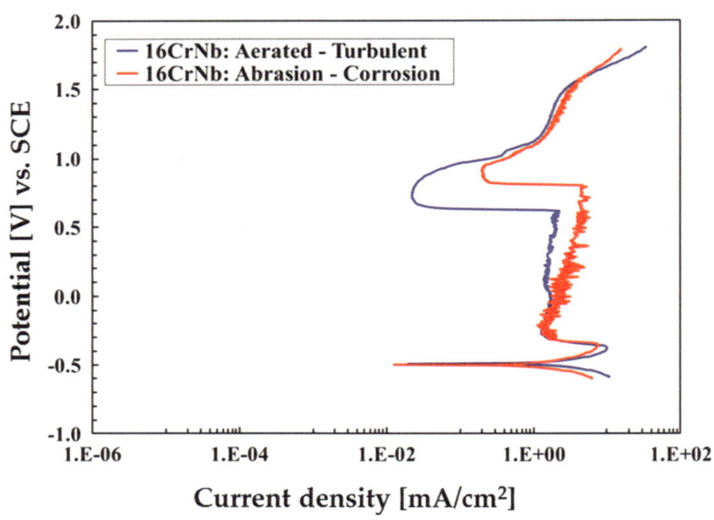

Figure 9. Typical potentiodynamic polarization curves for 16CrNb stainless steel [36].

- 11%wt Cr with (11CrTi) and without Ti (11Cr) stabilization;
- 16%wt Cr with (16CrNb) and without Nb (16Cr) stabilization;
- For comparative purposes, one austenitic stainless steel with 18%wt Cr and 8%wt Ni (18Cr8Ni) and one carbon steel with 0.15%wt C (ASTM A36).

Their chemical composition is presented in **Table 1**. Moreover, the specimens were thoroughly characterized in terms of mechanical properties and microstructure, details can be found in [36]. **Table 2** summarizes the main mechanical properties of the ferritic stainless steel.

The specimens were characterized in terms of corrosion, microabrasion and microabrasion-corrosion. Potentiodynamic curves were obtained for three test conditions. The first involved pure corrosion tests following the standard ASTM G5-94 [23]. The second involved modified corrosion tests in a turbulent and aerated medium. The third involved microabrasion corrosion tests.

The potentiodynamic curves obtained for the three different test conditions are exemplified in **Figure 10**. All the stainless steel specimens showed regions of cathodic behavior, anodic behavior, passive behavior and transpassive behavior. As expected, the carbon steel (A36) did not show a clear passive behavior (**Figure 10g**).

Comparing the standard corrosion tests with the corrosion tests using aerated conditions, it is possible to observe a substantial increase in the passivation current density (I_p) for all the materials tested. This suggests that the mechanical effect of the turbulence generated by pumping the solution rendered passivation more difficult.

This effect of aerated environments was observed for the corrosion of stainless steel in NaCl solutions by Qiao et al. [43] and by Le Bozec et al. [44] in saturated solutions of oxygen. The latter [44] found that under conditions of oxygen saturation, the anodic and cathodic reactions were accentuated. There is an increase in mass transport by oxygen, accelerating the corrosive process [45]. For the microabrasion-corrosion tests, I_p was further increased, attributable to the abrasion component, which removes the passive film.

Material	YS 0.2% (MPa)	TS (MPa)	El (%)	HV_{10} (MPa)
11Cr	323.8 ± 5.7	411.6 ± 2.0	37.7 ± 0.7	1510 ± 13
11CrTi	316.3 ± 4.1	400.1 ± 1.1	40.4 ± 1.9	1435 ± 11
16Cr	343.1 ± 3.9	488.3 ± 2.9	31.3 ± 1.8	1630 ± 29
16CrNb	336.2 ± 3.3	446.8 ± 1.7	35.9 ± 1.7	1514 ± 15
17CrTiNb	361.3 ± 1.4	483.2 ± 0.8	36.2 ± 0.2	1734 ± 15
18Cr8Ni	347.3 ± 9.4	706.2 ± 7.1	62.3 ± 1.2	1992 ± 9
A36	305.3 ± 4.2	444.8 ± 3.2	34.1 ± 1.0	1358 ± 11

Table 2. Mechanical properties of the specimens; YS = yield strength; TS = tensile strength; El = elongation; HV_{10} = Vickers hardness under 10 kgf normal load.

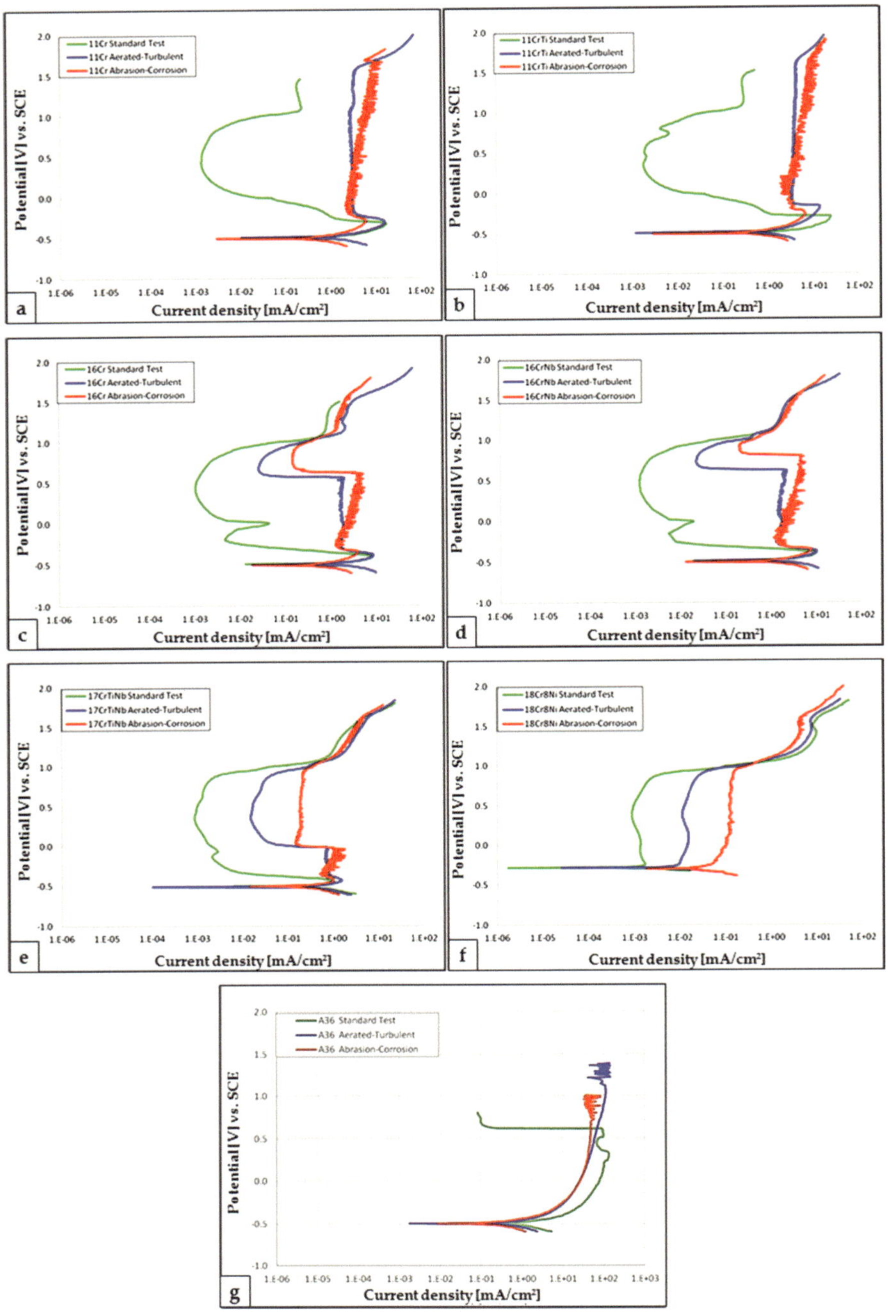

Figure 10. Polarization curves obtained for the three test conditions: (i) standard tests based on ASTM G5-94 (2004); (ii) aerated and turbulent corrosion conditions; and (iii) microabrasion-corrosion tests: (a) 11Cr, (b) 11CrTi, (c) 16Cr, (d) 16CrNb, (e) 17CrTiNb, (f) 18Cr8Ni, (g) A36.

On the other hand, additional interesting features could be observed for the polarization curves obtained under non-standard conditions. For steel with approximately 11% Cr (11Cr and 11CrTi, independently of stabilization), the potentiodynamic curves obtained in the microabrasion-corrosion condition did not show a regular passivation current density plateau; instead, I_p increased slightly but steadily with the potential (**Figure 10a** and **b**). Various authors [4, 40, 41] discuss this increase in passivation current density Ip as a result of the competition between the removal of the passive layer and repassivation. In fact, since it is not an effective passivation, it has been referred to as pseudo-passivation.

When the amount of Cr in the stainless steel increased to 16% (16Cr and 16CrNb, **Figure 10c** and **d**), the aerated tests presented a pseudo-passivation region for lower potentials (between around −400 mV and + 550 mV). After that, a secondary passivation led the current density to stabilize at lower values. Under microabrasion-corrosion conditions, in the pseudo-passive region, the current density again increased slightly with the potential. The secondary passivation still occurred, but at higher potentials. Increasing further the amount of Cr to 17% (**Figure 10e**), the pseudo-passive region and the secondary passivation region were still present for the aerated and microabrasion-corrosion conditions, but the potential for secondary passivation reduced, that is, the mechanical effects influenced less the electrochemical behavior of the material. Aerated conditions increased the passivation current, which was further increased under microabrasion-corrosion conditions.

The reference austenitic stainless steel (**Figure 10f**) with 18%Cr and 8%Ni did not present a pseudo-passive region for the non-standard tests. The current density remained constant during the passive plateau, but again, the agitation of the fluid increased the passivation current and agitation + rubbing increased it further.

The values of passivation current are summarized in **Figure 11**. For the lower Cr content (11%), the passivation currents in fact referred to a pseudo-passive behavior. First (**Figure 11**) shows that, as expected, an increase in the amount of Cr reduced the passivation current, for all the conditions tested. The increase in chromium content tends to accentuate the chemical adsorption through covalent coordinate bonds between the chromium and sulfur (S) atoms in the group of SO_4^{-2}. This fact is in agreement with molecular dynamic simulations by Diawara et al. [46], where the stability of the passive film grows from with Cr content, reaching its maximum at 20% Cr. It is also notorious, the increase in passivation current density I_p with the increase of the intensity of the mechanical event on the surface, first by agitation and then by abrasion.

Another factor that influences the increase of the passive current density in abrasion-corrosion tests is the application and/or increase of the normal force and plastic deformation on the surface of a body in passivation state [10, 42–44]. Ferritic stainless steel samples were subjected to fatigue by means of a servo-hydraulic "Schenck" machine at room temperature applying a load of 40 KN at frequencies of 0.1 and 0.5 Hz in environments of 1 M H_2SO_4 and 0.6 M NaCl [44]. In this test, potentiostatic conditions were imposed using a saturated silver electrode as reference and an exposed area of 3 × 7 mm^2. It was observed that current

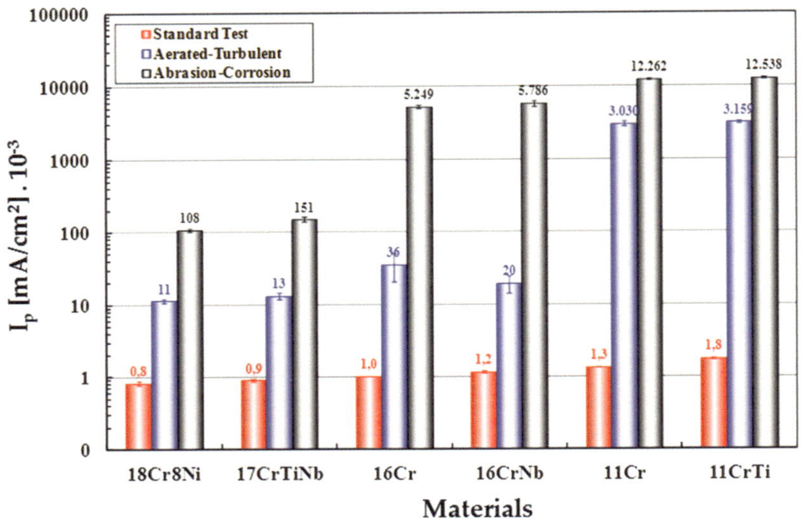

Figure 11. Results of passivation current density for the stainless steel specimens under the three conditions [47].

density increased with the rupture of the passive film due to plastic deformation, and it reduced again after the regeneration of the film on the virgin surface exposed to the corrosive environment. It is worth mentioning that the main difference between low and high frequencies is the time of repassivation of the exposed metal surface, that is, the rate of destruction of the passive film in relation to its generation and growth [44]. In the development of abrasion-corrosion maps [43], testing on a fixed-sphere microabrasion test rig using ultra-high molecular weight polyethylene ball with a suitable electrochemical cell in a physiological solution environment (0.9% NaCl and 10% FCS fetal calf serum), it was observed that for a potential in the 200-mV range, the system current doubled when the force changed from 2 to 3 N.

The normal force was varied during potentiodynamic abrasion-corrosion tests within the region where the imposed electrochemical conditions induced passivation [38]. The specimens were AISI 304 stainless steel, for an environment using abrasive particles of silica in 1 N H_2SO_4 solution and a zirconia ball. **Figure 12** shows a typical polarization curve obtained under these conditions.

The test started with an applied normal force of 1.42 N and during the test the force was decreased to 0.50 N. After 6 min, the load returned to 1.42 N. It is evident that when the force was reduced (from 1.5 to 0.5 N), the passivation current decreased, so the removal of the passive layer was more effective at higher loads [38]. The load variation also induced a change in friction coefficient (**Figure 13**).

In **Figure 13**, it is observed that when the normal force is reduced, the friction coefficient also reduces. The passive layer of the abrasive-corrosive process has lubricating properties, and with lower forces, the passive layer remains more constant and regenerates more efficiently.

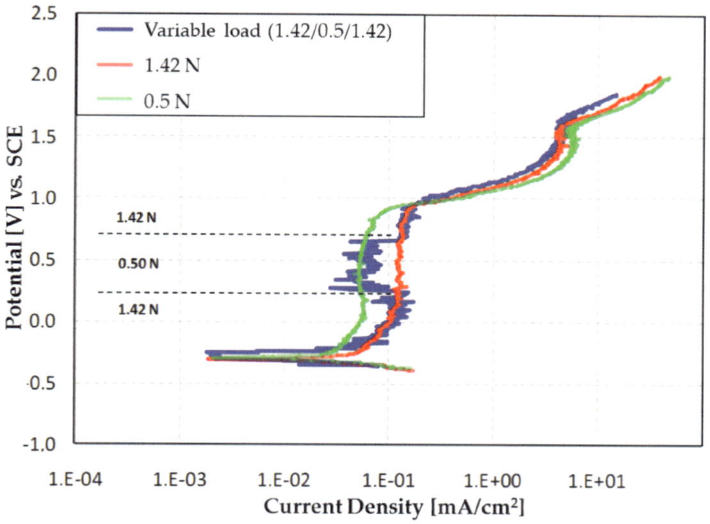

Figure 12. Polarization curves for an abrasion-corrosion test on AISI 304 with variable normal load [10].

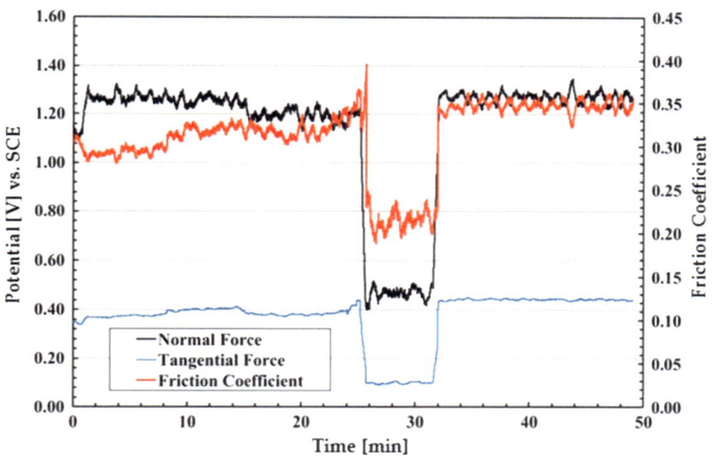

Figure 13. Evolution of contact forces for an abrasion-corrosion test on AISI 304 with variable normal load [10].

The formation of $FeSO_4$ in tribocorrosive systems containing sulfuric acid is the main factor responsible for the reduction of the friction coefficient [48]. Corrosion tests during reciprocal sliding of iron also showed a decrease in friction coefficient when sliding in H_2SO_4 solutions as compared to pure sliding tests using water [49]. In the presence of a H_2SO_4 solution as electrolyte, iron dissolution leads to the formation of $FeSO_4$ on the metallic surface, as indicated by the reaction shown in Eq. (1).

$$Fe_{(s)} + H_2SO_{4(aq)} \rightarrow FeSO_{4(aq)} + H_{2(g)} \tag{1}$$

To show the existence of the $FeSO_4$ film, a sample of ferritic stainless steel 410D (11Cr) was exposed to a 1 N H_2SO_4 solution for 1 hour, and then subjected to Fourier transform infrared spectroscopy (FTIR) analysis (**Figure 14**).

The FTIR spectrum identified humidity bands in the region between 3570 and 2940 cm^{-1} and 1650 cm^{-1}, and more importantly, it confirmed the formation of $FeSO_4$ in the region between 1168 and 1068 cm^{-1} [36].

To verify the kinetics response of the effects of corrosion on friction coefficient, two further test sequences were carried out (**Figure 15**).

In the first (**Figure 15a**), the abrasion-corrosion test was initiated using a water-based abrasive slurry, which was then changed to an abrasive slurry in 1 N H_2SO_4 solution. In the second (**Figure 15b**), the abrasion corrosion test started to use abrasive slurry in 1 N H_2SO_4 solution, which was changed to the abrasive slurry in water. Both curves show a short interval between the two situations, which corresponds to the time required to remove the load and change the slurry. When H_2SO_4 was added to the suspension (**Figure 15a**), the film appeared to form very rapidly, reducing the friction coefficient. In **Figure 15b**, abrasion removed the film, which was not restored in the absence of H_2SO_4, causing increased friction [36].

When comparing the friction coefficients measured during abrasion tests and abrasion-corrosion tests in 1 N H_2SO_4 environment (**Figure 16a**), the influence of the $FeSO_4$ film is evidenced.

The friction coefficient is two times higher in the abrasion tests when compared to the abrasion-corrosion tests. The reduction of friction coefficient under microabrasion-corrosion conditions decreases the amount of mechanical energy that is dissipated in the active interface in the form of friction [36]. This influence is also reflected in the wear coefficient (k). The wear coefficient is substantially lower (3–8 times lower) under abrasion-corrosion conditions when compared to pure abrasion conditions (**Figure 16b**). This behavior was also observed using NaCl solutions

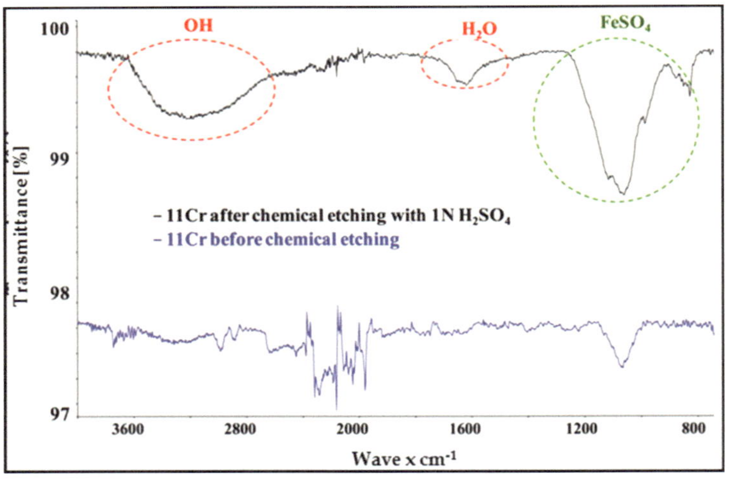

Figure 14. FTIR analysis of 11Cr ferritic stainless steel before and after immersion in 1 N H_2SO_4 solution [15].

http://dx.doi.org/10.5772/intechopen.81913

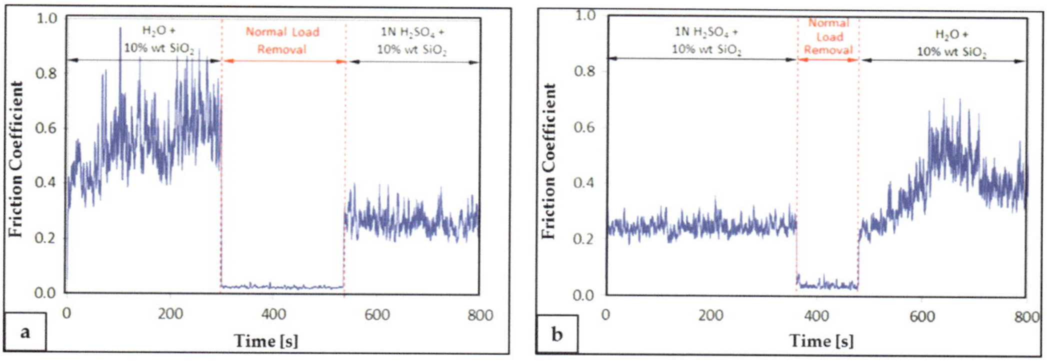

Figure 15. Friction coefficients measured during microabrasion-corrosion tests of 18Cr8Ni: (a) started with water slurry and replaced by 1 N H_2SO_4 slurry; (b) started with 1 N H_2SO_4 slurry and replaced by water slurry [36].

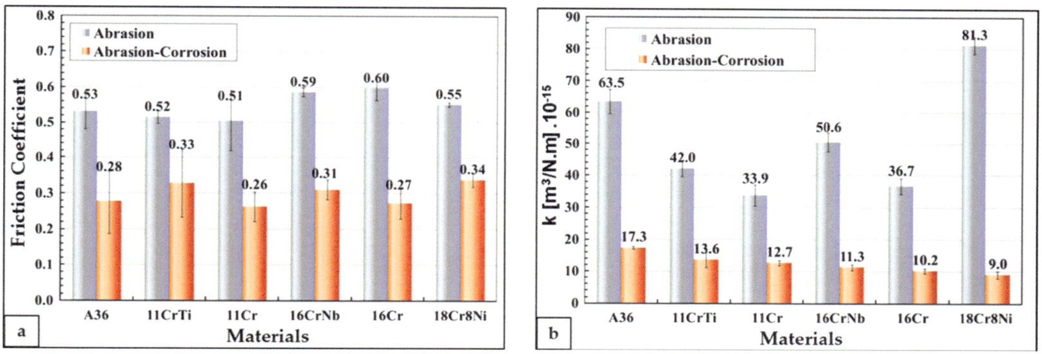

Figure 16. Evaluation of: (a) friction coefficient and (b) wear rate coefficient k, in pure microabrasive and microabrasive-corrosive environment [36].

for AISI 304 stainless steel and AISI 1045 carbon steel [37, 50]. This behavior is associated with the change in repassivation kinetics due to the presence of the corrosive medium and to the external application of a potential difference and the nature of the passive layer [36].

The comparative behavior between the different materials was very different from that observed under pure abrasion conditions, where the austenitic stainless steel had presented the worst performance. Under abrasion-corrosion conditions, the increase in chromium content resulted in a steady reduction of the wear coefficient. Higher Cr contents increase the stability of the passive film in more corrosive environments [28]. Under abrasion-corrosion conditions, the passive film stability seems to govern the behavior of stainless steel [7, 13]. The increase in Cr content could improve the film stability by either hindering depassivation or by accelerating repassivation [36]. The exact mechanism still needs to be elucidated and should be a niche of further research.

The lower wear rates are evidenced in the appearance of wear scars (**Figure 17**). For the two tests, abrasion and abrasion-corrosion arrangements, the predominance of the grooving mechanism is noted. This process occurs in the microscale abrasion test when a significant proportion of the particles slide at the interface producing a series of thin and parallel grooves

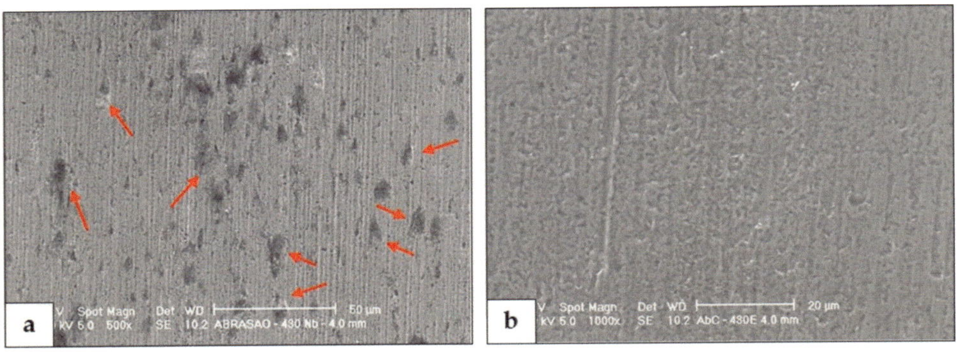

Figure 17. Analysis by SEM of the surfaces of 16CrNb ferritic stainless steel samples after wear testing: (a) microabrasion and (b) microabrasion-corrosion test [52].

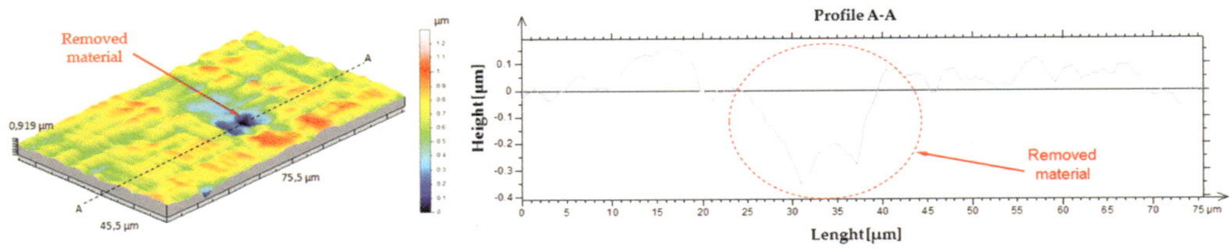

Figure 18. Topographic analysis of the abrasion wear craters evidencing large portions of material removal [17].

on the surface of the sample [51]. **Figure 17** shows that abrasion-corrosion leads to the formation of a significantly smoother surface, which results in less friction [38]. In this way, the smoother surface of abrasion-corrosion (**Figure 17b**) was observed in relation to microabrasion (**Figure 17a**), as a consequence of the low coefficient of friction [38].

The detachment of large portions of material is noteworthy for microabrasion tests (**Figure 17a**), indicated by arrows. Similar localized removal did not occur for the microabrasion-corrosion tests (**Figure 17b**), which showed lower wear coefficients than the carbon steel and the austenitic stainless steel. A 3D topographic assessment of the worn surfaces by laser interferometry (**Figure 18**) confirms that the regions indicated by arrows correspond indeed to areas with intense material removal.

The reduction of friction coefficient under abrasion-corrosion conditions decreases the amount of mechanical energy that is dissipated at the active interface in the form of friction. Vickers microhardness measurements were performed within the wear scars for all materials tested under conditions of pure microabrasion and abrasion-corrosion. The results are shown in **Figure 19**.

According to **Figure 19**, all materials showed an increase in hardness after testing as compared to the microhardness of the samples prior to testing. Strain hardening is a common phenomenon during metal abrasion and has been widely reported in the literature [53, 54]. However, stress hardening was less intense under abrasion-corrosion conditions than under abrasion conditions.

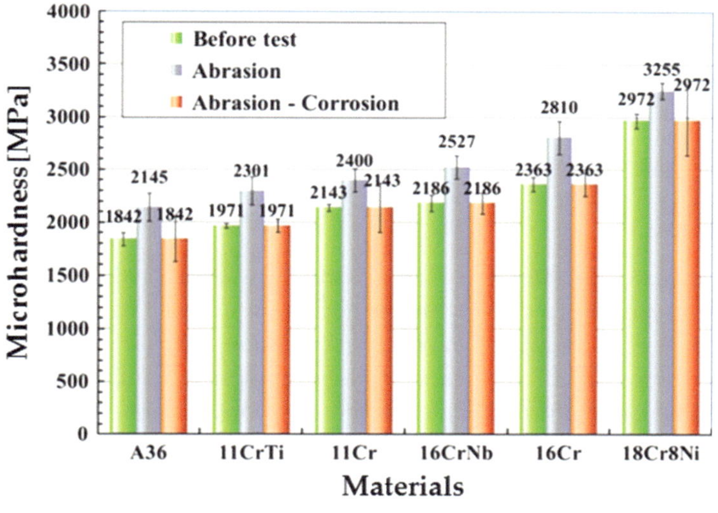

Figure 19. Micro hardness Vickers $HV_{0,05}$ measured within the wear scar before and after microabrasion and microabrasion-corrosion test [36].

Comparing pure microabrasion, which is mechanically dominated, with abrasion-corrosion, use of less energy in mechanical phenomena such as stress hardening was observed. This is in accordance with the lower values of friction coefficient measured under abrasion-corrosion than under pure abrasion, since abrasion-corrosion only needs friction energy for the mechanical removal of passive film, but not for the tribocorrosion of the active areas [36].

5. Conclusions

The main focus of this chapter was the introduction and performance assessment of ferritic stainless steel as an economical option, in particular, under abrasion-corrosion conditions. The comprehensive analysis for situations where pure corrosion, pure abrasion and simultaneous abrasion-corrosion occur allowed to conclude that:

- Particular emphasis was given to the initial stages of sugar cane processing: The ferritic stainless steel, despite its lowest price, showed wear resistance superior to that of more expensive austenitic stainless steel. This resulted in a further investigation of the performance of the ferritic stainless steel when compared to low-carbon steel in a pilot study in three industrial plants. With the successful results obtained with the pilot substitution of some parts that were conventionally manufactured using carbon steel by similar parts in stainless steel, which started in 2003, this substitution increased, so that since 2007 almost all parts in the initial stages of sugar cane processing in the main Brazilian industrial plants have been manufactured using 11Cr stainless steel.
- Both crystallographic texture and Nb/Ti stabilization exerted a strong influence on the corrosion of ferritic stainless steel. Local corrosion (pitting) exhibited an anisotropic behavior and

was influenced by the crystallographic texture, particularly concerning the pitting potential. This anisotropic behavior showed that the presence of the Goss shear texture (oriented along <100>) had a positive influence on the corrosion resistance. Ti and Nb stabilization helped to protect against intergranular corrosion, as the grain borders become less reactive.

- A substantial difference in wear rate, friction coefficient and passive current density was observed for tests performed under different configurations (i.e., corrosion, abrasion and abrasion-corrosion).
- Although the austenitic stainless steel (18Cr8Ni), with the highest Cr content among the materials tested, showed the worst performance under pure abrasion conditions, it presented the best performance in the microabrasion-corrosion tests. Under abrasion-corrosion conditions, the tribological performance improved with the increase in Cr content. The mechanical effects of turbulence and abrasion accelerated the corrosion process, mainly evidenced by an increase in the passivation current density.
- For the abrasion-corrosion tests, a negative synergy was observed. For this configuration, an increase in contact force produced an increase in passive current density. Tests with variable normal force gave a clear indication that this behavior is reversible. The test rig allowed the measurement of friction coefficients during microabrasion tests. This allowed to detect a reduction in friction coefficient during abrasion-corrosion conditions when compared with pure abrasion tests. Friction reduction was attributed to the presence of $FeSO_4$ as a corrosion product, which was confirmed by FTIR analysis.
- Microhardness Vickers measurements inside the wear scars showed a less intense strain hardening under abrasion corrosion conditions than under pure abrasion conditions. This is probably due to the reduction in the energy dissipated as friction in the contact and therefore available for plastic deformation.

Author details

Wilian S. Labiapari[1,2], Miguel A. N. Ardila[1], Henara L. Costa[1,3] and José Daniel B. de Mello[1]*

*Address all correspondence to: ltm-demello@ufu.br

1 Federal University of Uberlândia, Uberlândia, Brazil

2 Aperam South America, Timóteo, Brazil

3 Federal University of Rio Grande, Rio Grande, Brazil

References

[1] Stack MM, Chi K. Mapping sliding wear of steels in aqueous conditions. Wear. 2003;**255**:456-465

[2] Stack MM, Jawan H, Mathew MT. On the construction of micro-abrasion maps for a steel/polymer couple in corrosive environments. Tribology International. 2005;**38**:848-856

[3] Labiapari WS, Alcântara CM, Costa HL, de Mello JDB. Stainless steel as an antiwear material for the bio-fuel industry. Wear. 2013;**302**:1536-1545

[4] Wood RJK, Sun D, Thakare MR, de Frutos Rozas A, Wharton JA. Interpretation of electrochemical measurements made during micro-scale abrasion-corrosion. Tribology International. 2010;**43**:1218-1227

[5] Landolt D, Mischler S, Stemp M, Barril S. Third body effects and material fluxes in tribo-corrosion systems involving a sliding contact. Wear. 2004;**256**:517-524

[6] Bello JO, Wood RJK, Wharton JA. Synergistic effects of micro-abrasion–corrosion of UNS S30403, S31603 and S32760 stainless steels. Wear. 2007;**263**:149-159

[7] Gant AJ, Gee MG, May AT. The evaluation of tribo-corrosion synergy for WC–Co hard-metals in low stress abrasion. Wear. 2004;**256**:500-516

[8] Sun D, Wharton JA, Wood RJK. Micro-abrasion mechanisms of cast CoCrMo in simulated body fluids. Wear. 2009;**267**:1845-1855

[9] Thakare MR, Wharton JA, Wood RJK, Menger C. Exposure effects of strong alkaline conditions on the microscale abrasion–corrosion of D-gun sprayed WC–10Co–4Cr coating. Tribology International. 2008;**41**:629-639

[10] ANFAVEA, Monthly Report, Associacão Nacional dos Fabricantes de Veículos Automotores Brazil; 2012

[11] Volci GA. Comportamento tribológico do anel de primeiro canalete em motores operando em sistemas flex fuel, Mechanical Engineering. Brazil: UFPR; 2007. p. 116

[12] Cheng YF, Bullerwell J, Steward FR. Electrochemical investigation of the corrosion behavior of chromium-modified carbon steels in water. Electrochimica Acta. 2003;**48**:1521-1530

[13] Dobbelaar JAL, Herman ECM, de Wit JHW. The corrosion behaviour of iron-chromium alloys in 0.5 M sulphuric acid. Corrosion Science. 1992;**33**:765-778

[14] Huh MY, Lee JH, Park SH, Engler O, Raabe D. Effect of through-thickness macro and micro-texture gradients on ridging of 17% Cr ferritic stainless steel sheet. Steel Research International. 2005;**76**:797-806

[15] Hutchinson B. Deformation microstructures and textures in steels. Philosophical Transactions of the Royal Society of London A: Mathematical, Physical and Engineering Sciences. 1999;**357**:1471-1485

[16] Kumar BR, Singh AK, Das S, Bhattacharya DK. Cold rolling texture in AISI 304 stainless steel. Materials Science and Engineering A. 2004;**364**:132-139

[17] Raabe D, Lücke K. Influence of particles on recrystallization textures of ferritic stainless steels. Steel Research. 1992;**63**:457-464

[18] Raabe D, Lücke K. Texture and microstructure of hot rolled steel. Scripta Metallurgica et Materialia. 1992;**26**:1221-1226

[19] Raabe D, Lüucke K. Textures of ferritic stainless steels. Materials Science and Technology. 1993;**9**:302-312

[20] Raabe D, Reher F, Hölscher M, Lücke K. Textures of strip cast Fe16% Cr. Scripta Metallurgica et Materialia. 1993;**29**:113-116

[21] Ardila MAN, Labiapari WS, De Mello JDB. The influence of crystallographic texture and niobium stabilisation on the corrosion resistance of ferritic stainless steel. Materials Research. 2017;**20**:576-583

[22] Siqueira RP, Sandim HRZ, Oliveira TR, Raabe D. Composition and orientation effects on the final recrystallization texture of coarse-grained Nb-containing AISI 430 ferritic stainless steels. Materials Science and Engineering A. 2011;**528**:3513-3519

[23] ASTM. Standard Test Method for Conducting Potentiodynamic Polarization Resistance Measurements. West Conshohocken, PA: ASTM International; 1991. pp. G59-G97

[24] Huang C, Zou B, Liu Y, Zhang S, Huang C, Li S. Study on friction characterization and wear-resistance properties of Si3N4 ceramic sliding against different high-temperature alloys. Ceramics International. 2016;**42**:17210-17221

[25] Ma L, Hu S, Shen J, Han J, Zhu Z. Effects of Cr content on the microstructure and properties of 26Cr–3.5Mo–2Ni and 29Cr–3.5Mo–2Ni super ferritic stainless steels. Journal of Materials Science and Technology. 2016;**32**:552-560

[26] Hamdy AS, El-Shenawy E, El-Bitar T. The corrosion behavior of niobium bearing cold deformed austenitic stainless steels in 3.5% NaCl solution. Materials Letters. 2007;**61**:2827-2832

[27] Kiminami CS, Souza CAC, Bonavina LF, de Andrade Lima LRP, Suriñach S, Baró MD, et al. Partial crystallization and corrosion resistance of amorphous Fe-Cr-M-B (M=Mo, Nb) alloys. Journal of Non-Crystalline Solids. 2010;**356**:2651-2657

[28] Koga GY, Nogueira RP, Roche V, Yavari AR, Melle AK, Gallego J, et al. Corrosion properties of Fe–Cr–Nb–B amorphous alloys and coatings. Surface and Coatings Technology. 2014;**254**:238-243

[29] Sousa CAC, Kuri SE. Relationship between niobium content and pitting corrosion resistance in ferritic stainless steels. Materials Letters. 1995;**25**:57-60

[30] Davis BW, Moran PJ, Natishan PM. Metastable pitting behavior of aluminum single crystals. Corrosion Science. 2000;**42**:2187-2192

[31] Park SY, Kim JH, Lee MH, Jeong YH. Effects of the microstructure and alloying elements on the iodine-induced stress-corrosion cracking behavior of nuclear fuel claddings. Journal of Nuclear Materials. 2008;**376**:98-107

[32] Shahryari A, Szpunar JA, Omanovic S. The influence of crystallographic orientation distribution on 316LVM stainless steel pitting behavior. Corrosion Science. 2009;**51**:677-682

[33] Kruger J. Influence of crystallographic orientation on the pitting of iron in distilled water. Journal of the Electrochemical Society. 1959;**106**:736

[34] Guo WY, Sun J, Wu JS. Effect of deformation on corrosion behavior of Ti–23Nb–0.7Ta–2Zr–O alloy. Materials Characterization. 2009;**60**:173-177

[35] Costa HL, Ardila MAN, Labiapari WS, Silva WM, de Mello JDB. Effect of surface topography on the dynamics of the abrasive particles during micro-abrasion. Wear. 2015;**324-325**:129-139

[36] Labiapari WS, Ardila MAN, Costa HL, de Mello JDB. Micro abrasion-corrosion of ferritic stainless steels. Wear. 2017;**376-377**(Part B):1298-1306

[37] Oltra R, Chapey B, Renaud L. Abrasion-corrosion studies of passive stainless steels in acidic media: Combination of acoustic emission and electrochemical techniques. Wear. 1995;**186-187**:533-541

[38] Santos MB, Labiapari WS, Ardila MAN, da Silva WM, de Mello JDB. Abrasion–corrosion: New insights from force measurements. Wear. 2015;**332-333**:1206-1214

[39] Sinnett-Jones PE, Wharton JA, Wood RJK. Micro-abrasion–corrosion of a CoCrMo alloy in simulated artificial hip joint environments. Wear. 2005;**259**:898-909

[40] Stack MM, Zhou S, Newman RC. Identification of transitions in erosion-corrosion regimes in aqueous environments. Wear. 1995;**186-187**:523-532

[41] Ferrer F, Idrissi H, Mazille H, Fleischmann P, Labeeuw P. A study of abrasion–corrosion of AISI 304L austenitic stainless steel in saline solution using acoustic emission technique. NDT and E International. 2000;**33**:363-371

[42] Mischler S, Rosset E, Stachowiak GW, Landolt D. Effect of sulphuric acid concentration on the rate of tribocorrosion of iron. Wear. 1993;**167**:101-108

[43] Qiao YX, Zheng YG, Okafor PC, Ke W. Electrochemical behaviour of high nitrogen bearing stainless steel in acidic chloride solution: Effects of oxygen, acid concentration and surface roughness. Electrochimica Acta. 2009;**54**:2298-2304

[44] Le Bozec N, Compère C, L'Her M, Laouenan A, Costa D, Marcus P. Influence of stainless steel surface treatment on the oxygen reduction reaction in seawater. Corrosion Science. 2001;**43**:765-786

[45] Freguia S, Rabaey K, Yuan Z, Keller J. Non-catalyzed cathodic oxygen reduction at graphite granules in microbial fuel cells. Electrochimica Acta. 2007;**53**:598-603

[46] Diawara B, Beh YA, Marcus P. Atomistic Simulation of the Passivation of Iron-Chromium Alloys Using Calculated Local Diffusion Activation Barriers; 2006

[47] Labiapari WS, Ardila MAN, Costa HL, de Mello JDB. Mechanical Effects on the Corrosion Resistance of Ferritic Stainless Steels During Microabrasion-Corrosion, Wear of Materials (WOM), Wear, Miami; 2019

[48] Kolesnichenko LF, Trushko PV. Formation of boundary layers during friction in the presence of sulfur. Soviet Powder Metallurgy and Metal Ceramics. 1970;**9**:993-997

[49] Miyoshi K. Studies of mechanochemical interactions in the tribological behavior of materials. Surface and Coatings Technology. 1990;**43**:799-812

[50] Bateni MR, Szpunar JA, Wang X, Li DY. The effect of wear and corrosion on internal crystalline texture of carbon steel and stainless steel. Wear. 2005;**259**:400-404

[51] Trezona RI, Allsopp DN, Hutchings IM. Transitions between two-body and three-body abrasive wear: Influence of test conditions in the microscale abrasive wear test. Wear. 1999;**225-229**(Part 1):205-214

[52] Labiapari WS. Abrasão-corrosão em aços inoxidáveis ferríticos. Uberlândia: Programa de Pós-graduação em Engenharia Mecânica, Universidade Federal de Uberlândia (UFU); 2015. p. 211

[53] Hutchings IM. Tribology: Friction and Wear of Engineering Materials. London: Edward Arnold; 1992

[54] Zum Gahr KH. Microstructure and Wear of Materials. Amsterdam: Elsevier Science; 1987

Microstructure-Tailored Stainless Steels with High Mechanical Performance at Elevated Temperature

Kamran Saeidi and Farid Akhtar

Additional information is available at the end of the chapter

http://dx.doi.org/10.5772/intechopen.80468

Abstract

Stainless steels are widely used in chemical, structural and automotive applications due to their high room-temperature mechanical properties, toughness, corrosion resistance and low cost. However, tendency and rise in industrial demands for components to be used at high temperature with good mechanical performance and corrosion resistance limit their usage in many applications and narrow down their service criteria. Tailoring the microstructure, tuning the chemistry, adjusting the phase composition and introducing a dense 3D network of dislocations can tailor and develop stainless steels with high performance for extreme conditions, such as elevated temperatures. In this chapter, the effect of the microstructure of additively manufactured and thermo-mechanically processed stainless steels on the high temperature mechanical performance is discussed and a comparison is made with conventional steels. Moreover, new mechanisms are introduced and discussed co-relating the microstructure and properties.

Keywords: stainless steel, selective laser melting, microstructure tailoring, high temperature, mechanical properties

1. Introduction

Additive manufacturing (AM) has been developed from prototyping to near net shape and end products which can challenge and compete with the traditional manufacturing methods such as casting, pressing and sintering. The components of materials, such as stainless steel, are built up with unique structures at various length scales spanning from macro down to nano metric levels which are different from materials produced by conventional methods. The AM technique makes it possible to assemble localized tuneable structures, tailor the chemistry and microstructure and advance the mechanical, physical and corrosion properties to produce customized materials. The tailoring and manipulation of the materials chemistry and the microstructure not only improves the room temperature mechanical properties of stainless steels but provides opportunities to design stainless steels for elevated temperatures and for relatively harsh environments. Industrial demands show that the stainless steels are of interest for high temperature applications nowadays since these materials can compete in price with other materials and therefore be more efficient and economical. This chapter introduces AM manufactured stainless steels (316L, 2507 and 420) and their post processing to tailor microstructures to achieve advanced mechanical properties both at room temperature and elevated temperatures and make comparisons with traditionally manufactured stainless steels. The outcome of this work is undoubtedly of interest for engineers and scientists.

2. Additive manufacturing

Additive manufacturing is a flexible technique that can manufacture three-dimensional (3D) solid objects of virtually any geometry from digital design data based on computer aided design and computer aided manufacturing (CAD/CAM). The AM technique builds up objects in a bottom-up approach, i.e. by adding material layer by layer on top of each other and consolidating it using a high energy heating source [1]. Early use of AM in the form of rapid prototyping focused on pre-production and nowadays AM is being used to fabricate end-use products. AM can be used to process materials such as plastics, polymers, metals, alloys, ceramics and human tissue. Based on the raw materials used, AM is categorized into three main groups: (I) solid-based, such as laminated object manufacturing (LOM) [2] where cut to shape laminates are bonded together; (II) liquid-based, such as fused deposition modeling (FDM) [3] Stereo-lithography (SLA) where a photosensitive polymer resin is cured layer by layer using a scanning ultraviolet laser beam [4, 5]; and (III) powder-based, such as selective laser sintering (SLS) or selective laser melting (SLM) [6–8] where layers are created by spreading the powder on the building substrate and melting the powder particles bonding them together using a high energy laser source (see **Figure 1**), and electron beam source, i.e., electron beam melting (EBM) [9–11] and three-dimensional printing (3DP) [12, 13] where layers are created from powders onto which a traversing ink jet head prints a bonding agent. LOM, FDM, SLA and 3DP can produce metal parts in an "indirect" approach in which a binder is used to bond precursor particles forming 3D parts and post-processing is required after the AM process. SLS/SLM, EBM are "direct" ways in which the precursor particles are partially melted and/or fully melted during the AM process to make the final part directly. In the following sections, the SLM processing of stainless steels and post processing to tailor structure and properties are discussed.

SLM is used to fabricate metallic prototypes, components and counterparts in various industrial applications such as aerospace, defense, spacecraft, automobile and the biomedical industries. **Figure 2** shows examples of products built by SLM technology.

For SLM fabrication of materials two requirements have to be met:

- **Powder precursor:** the powder should be provided as gas atomized spherical granules with size between 25 and 150 μm. The powder should have a good flowability to ensure

http://dx.doi.org/10.5772/intechopen.80468

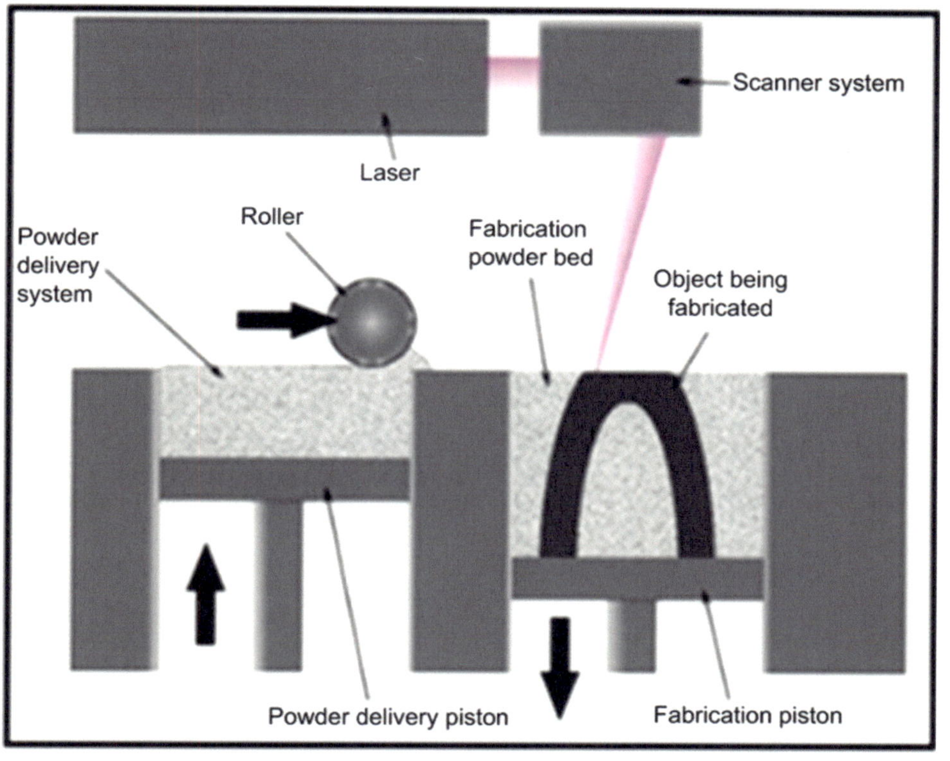

Figure 1. Schematic of the selective laser melting process (http://blog.trendonix.com/3d-printing-selective-laser-sintering).

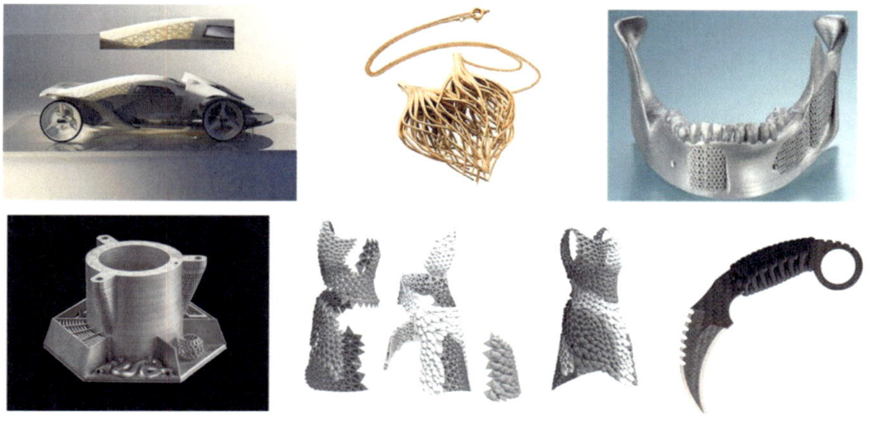

Figure 2. Example of products in different application fields produced by SLM technology (courtesy of Saeidi [14]).

unified spreading of the powder by the re-coater blade on the building substrate and an acceptable apparent density to avoid sinking of the powder.

- **Optimization of the processing parameters:** laser processing parameters are optimized to ensure a solid and decent part. The main parameters are laser power (P), scan speed (v), focal spot diameter (2a), line spacing (L) and layer thickness (t). Because of strong association between the laser power and scan speed a suitable way of expressing their influences on powder interaction is in the form of a compound variable (Eq.1) which is referred to as the laser energy density (LED) [1].

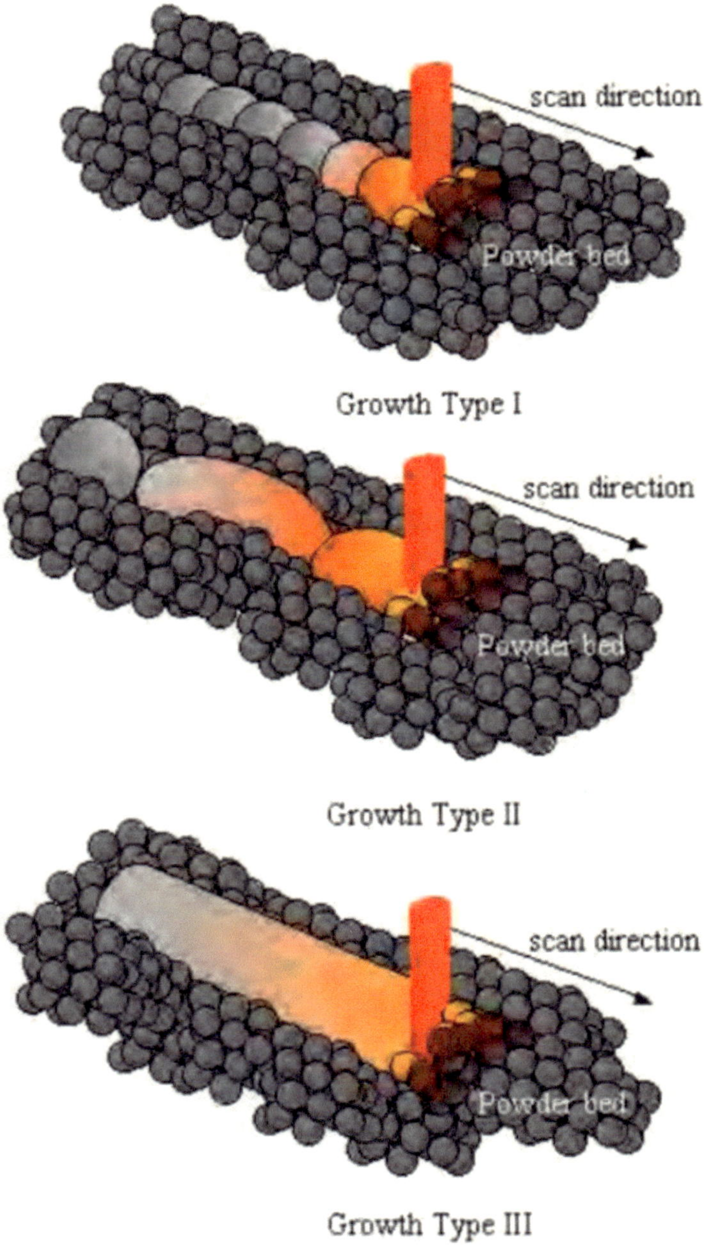

Figure 3. Three types of melt pools with growth type I (balled melt pool) type II (discontinuous and fragmented melt pools) and type III (continuous melt pool columns) (courtesy of Beng [15]).

$$LED = P/2a.V \quad (1)$$

During SLM, upon exposure of powder with the laser beam momentarily melt droplets are created and due to the moving beam elongated melt pools are created. These melt pools can be regarded as small casting. The shape of the melt pools can be controlled by the surface tension of the melt. The solidification of the melt results in the formation of microstructure within the melt pools, thus the shape and size of the melt pool are controlled and adjusted [15]. The size and shape of melt pools can be adjusted by controlling the laser processing parameters. Three different melt pool types, (I) balled melt pools; (II) discontinuous and fragmented melt pools; and (III) continuous melt pools have been identified in laser processing of materials, as shown in **Figure 3**. The fast cooling rate (10^5 KS^{-1}) for metallic materials and the confined melt pool results in unique microstructures [14].

2.1. Fabrication of stainless steels using SLM

Stainless steels are one of the most widely investigated materials for SLM [16] and based on the laser processing parameters and LED discussed above, SLM can fabricate dense [17, 18] and porous stainless steels used in aerospace, automotive, fuel cells [19] and biomedical applications [20, 21]. In this chapter, fabrication of stainless steel grades such as austenitic 316L, duplex 2507 and martensitic 420 grades with SLM technique and post processing of the stainless steels will be discussed and their influence on the microstructure and mechanical properties will be explained.

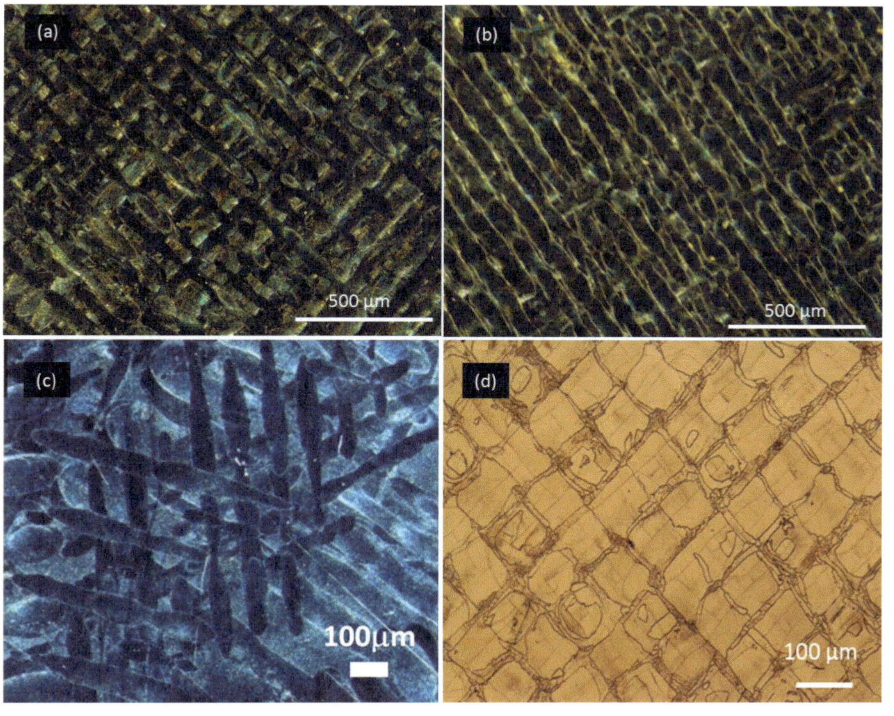

Figure 4. Different melt pool patterns caused by different scanning strategies during SLM of (a and b) 316L austenitic stainless steel, (c) CrCoMo dental alloy, (d) 2507 duplex stainless steel (figures have not been published elsewhere).

2.2. Microstructure hierarchies and heterogeneities in SLM stainless steel components

Hierarchical assembly of materials is one of the most promising yet challenging aspects of nanoscience [22–24]. Although synthesizing hierarchical materials, in terms of complexity and precision, is still very primitive, advanced progress has already been made towards this field. Recently, hierarchical structures are processed in materials, e.g., Ni-base super alloys processed by rapid solidification of the melt [25, 26]. Hierarchies in laser induced coating of Ti6Al4V alloys, in which a multi-scale textured zirconia coating was applied on titanium alloy using a pulsed laser, has been reported [27]. Lui et al. [28] synthesized W-Cr alloys with hierarchical structure using combustion synthesis. However, hierarchies in stainless steel have been reported recently [18] in the microstructure of selective laser melted steels. During SLM

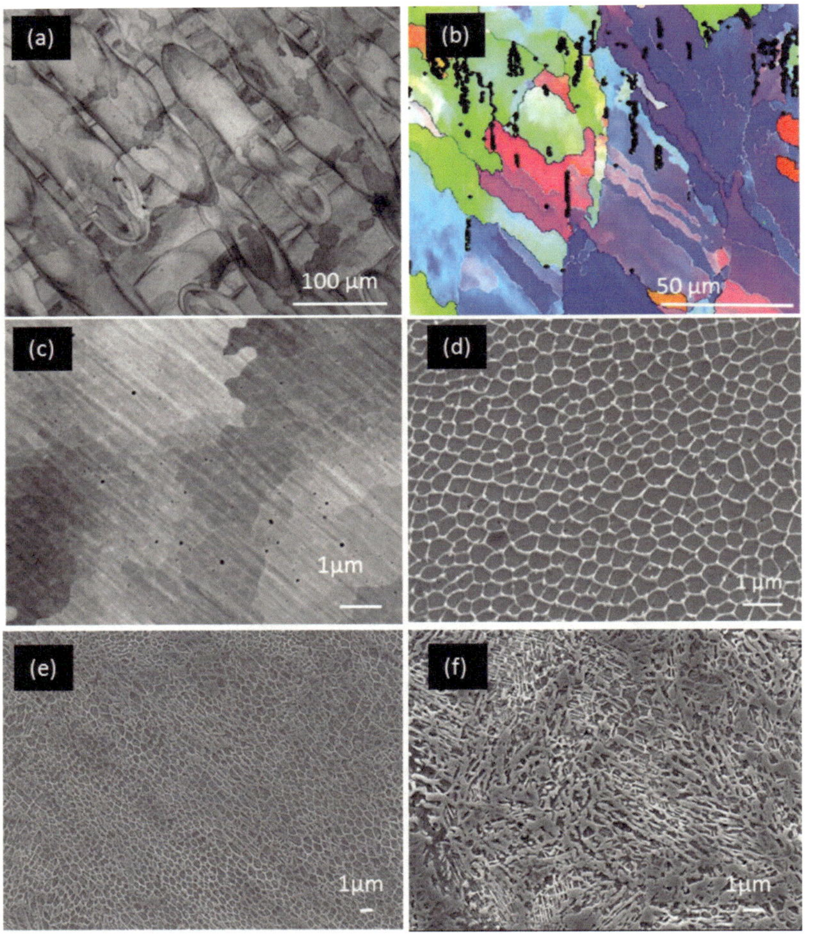

Figure 5. (a) Melt pools in macroscopic length scale, (b) EBSD image of SLM 316L stainless steel showing elongated grains inside the melt pools at microscopic length scale, (c) SEM image showing existence of a substructure inside the micro grains, (d) SEM image showing the substructure has a cellular morphology with size around 1 µm, (e) the cellular substructure in 420 martensitic stainless steel, (f) nano metric martensite needles inside the cells and across the cells in 420 steel (figures have not been published elsewhere).

of stainless steels network of melt pools are created at the macroscopic levels. The size of the melt pools is determined by the line spacing. The directions of melt pools and how they are built can be adjusted according to the laser scanning pattern. Changing or rotating the scanning pattern could generate complex network of the melt pools in stainless steels. **Figure 4** shows different scanning patterns for SLM 316L, CrCoMo dental alloy and SLM 2507.

During the SLM process, upon rapid solidification in the small volume of melt pool various crystallization processes can take place at microscopic level (**Figure 5a**). Inside the melt pools grains are formed as seen in from the EBSD orientation mapping showing elongated grains (**Figure 5b**). At the sub-micron scale and nano levels inside the grains a substructure is seen which is sown in **Figure 5c**. The substructure has cellular morphology with cells around 1 μm (**Figure 5d**). Hierarchy in laser melted 420 martensitic stainless steel is depicted in **Figure 5e** and **f** where cellular structure in micron scale (**Figure 5e**) and colonies of martensitic needles with different orientation in submicron scale (**Figure 5f**) are seen.

Hierarchical microstructure heterogeneities at two length scales macro and micro are present in the laser melted stainless steels. The microstructure heterogeneities can be categorized to grain morphology, grains size and orientation. **Figure 6a** shows existence of different

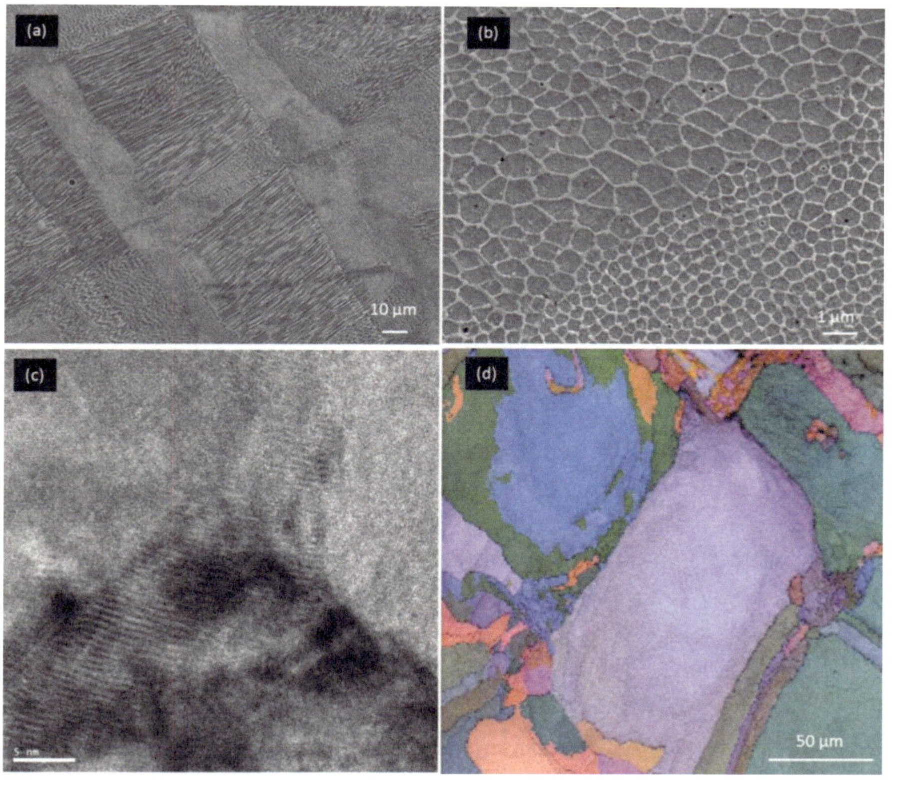

Figure 6. SEM images of SLM 316L showing (a) different grain morphology existing inside the melt pools, (b) cells with different sizes, (c) HRTEM image from SLM 316 L indicating different orientation of subgrains inside bigger grains, (d) EBSD images from SLM 2507 stainless steel confirming existence of smaller grains in between the tesserae and inside each tesserae (figures have not been published elsewhere.).

microstructure patterns, such as hexagonal cells, columnar and band structure. **Figure 6** illustrates these heterogeneities clearly. **Figure 6a** shows existence of different microstructure patterns, such as hexagonal cells, columnar and band structure. In **Figure 6b**, cells with different sizes are seen. As shown by the HRTEM in **Figure 6c**, the cells next to each other coexisting in a bigger grain have different orientations.

EBSD image reveals the existence of different grain size and orientation in duplex 2507 stainless steel processed by laser melting (**Figure 6d**). Morris et al. [29] proved the cellular substructure is in fact subgrains with low angle grain boundaries (LAGBs) that are confined in the bigger grains with high angle grain boundaries. The measured misorientation of two adjacent cells (subgrains) was around two degrees, further supporting their claim.

3. Mechanical properties of stainless steels

3.1. Mechanical performance and microstructural studies of SLM stainless steels at room temperature (25°C)

Stainless steels are primarily utilized because of their corrosion resistance, low cost and room temperature mechanical properties. The properties of stainless steels result from both its chemical composition, microstructure characteristics and its method of manufacture, including processing during fabrication. This section addresses room temperature mechanical properties of stainless steel grades (316L, 2507, 420) manufactured by SLM and a comparison is made with conventional steels. **Figure 7** shows the tensile test curve of SLM 316L at room temperature (a) and the corresponding micrograph of fracture surface (b).

The mechanical properties are listed in **Table 1**. As seen, the SLM 316L shows yield strength of 456 MPa and tensile strength of 700 MPa and microhardness of 325 Hv. These properties are higher than all conventional 300 series steels with yield strength between 240 and 300 MPa, tensile strength between 500 and 650 MPa and microhardness in the range of 140–180 Hv [30]. The ductility of the SLM 316L however is similar to conventional 316L. Despite having a high ductility, one major drawback of conventional 316L stainless steel is its low yield strength (~250–300 MPa) [31]. Many traditional metallurgical routes for strengthening 316L (for example, cold rolling, forging) inevitably results in a severe drop in ductility. To date, despite decades of studies the strategies to achieve both high strength and ductility remain rare. However, the mechanical strength of the SLM 316L is superior without sacrificial loss of ductility. Therefore, the SLM fabrication can overcome the strength–ductility trade-off that exists ubiquitously in pure metals and alloys [29].

The underlying mechanism breaking the strength-ductility trade has been investigated by detailed characterization of SLM 316 stainless steel (**Figure 8**). The SLM 316L possesses hierarchically heterogeneous microstructure with fine micron sized cellular subgrains with one or more orientation confined by macro grains (**Figure 8a**). The subgrains are formed by enrichment of elements such as Cr and Mo investigated by STEM and elemental mapping

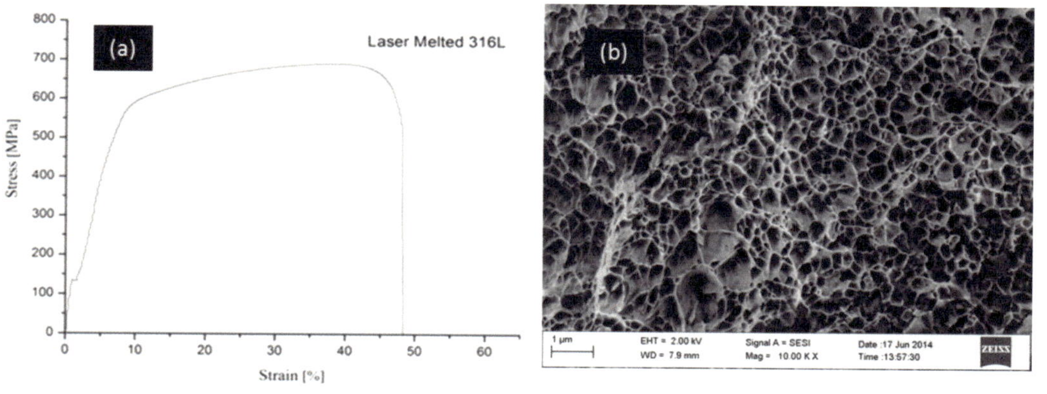

Figure 7. (a) Tensile test curve obtained from SLM 316 L at room temperature, (b) scanning electron micrograph of the fracture surface showing ductile dimples with size less than 1 µm (figures have not been published elsewhere).

Steel Type	Yield strength (MPa)	Tensile strength (MPa)	Elongation (%)	Hardness (Hv)
Austenitic	240-300	500-650	40-45	140-180
Ferritic	270-450	420-750	15-13	250-400
Martensitic	400-800	500-1100	10-15	280-560
Duplex	400-550	700-1000	20-25	270-310
SLM 316L[a]	456	703	45	325

Table 1. Mechanical properties of stainless steel types and comparison with SLM 316 L.

(**Figure 8b**). Movement of Mo with a large atomic radius around 1.99 Å inside the steel matrix generates lots of stress and forms dislocation loops (**Figure 8c**) [17]. Dislocation interactions with each other and with low angle grain boundaries (**Figure 8d** and **e**), formed by the subgrains, increases the mechanical strength whilst presence of slip planes, deformation twins [29] and different subgrain size avoids ductility loss. The superior properties arise from the collective effects of hierarchically heterogeneous microstructures, including solidification cellular structure, LAGBs, and dislocations.

Room temperature mechanical properties of SLM duplex 2507 are listed in **Table 2**. The material exhibits a superior mechanical strength with yield strength of 1200 MPa, tensile strength of 1320 MPa and microhardness of 450 Hv which are almost two times higher than all grades of ferritic and duplex stainless steel reported by AutoCompu [30]. However, the elongation of the studied SLM 2507 is two times lower than conventional duplex steels (10% for SLM 2507 compared to 20–25% for conventional duplex stainless steels).

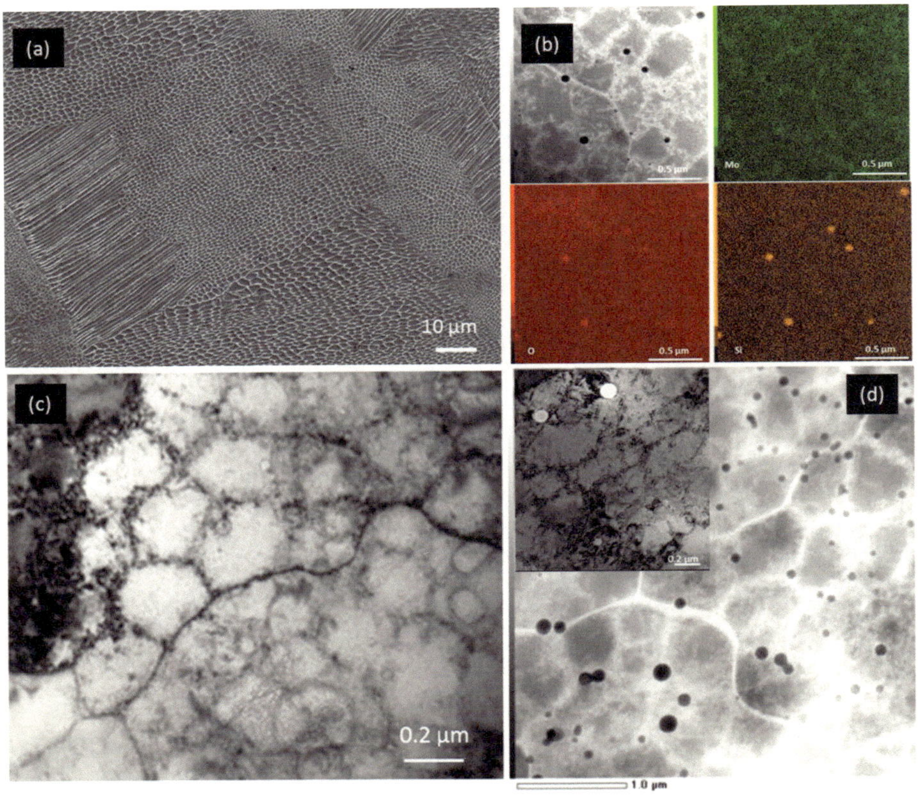

Figure 8. (a) SLM 316L possesses hierarchically heterogeneous microstructure with fine micron sized cellular subgrains with one or more orientation confined by macro grains, (b) STEM and elemental mapping indicating subgrains are formed by enrichment of elements such as Cr and Mo (copyright obtained; *DOI: 10.1039/C4RA16721J*), (c) bright field TEM image showing dislocation loops formed around the cell boundaries (copyright obtained; DOI: 10.1016/j.msea.2014.12.018), (d) oxide nano inclusions strengthening the 316L steel by pinning the dislocation movement as seen from the inset.

Type	Tensile strength (MPa)	Yield strength (MPa)	Microhardness (HV)
SLM Duplex	1321±48	1214±43	450
Duplex SS	600-800	400-550	200-250
Ferritic SS	400-600	250-400	135-175

Table 2. Comparison of room temperature mechanical properties of SLM duplex with conventional duplex and ferritic stainless steel.

The reason behind superior mechanical strength is combination of macro mosaic texture containing grains shown by EBSD image (**Figure 9a** and **b**) very dense 3D network of dislocation forming loops with diameter of 200 nm (**Figure 9c** and **d**), nitrogen enriched zones and formation of hard and brittle nitrides (**Figure 9e** and **f**).

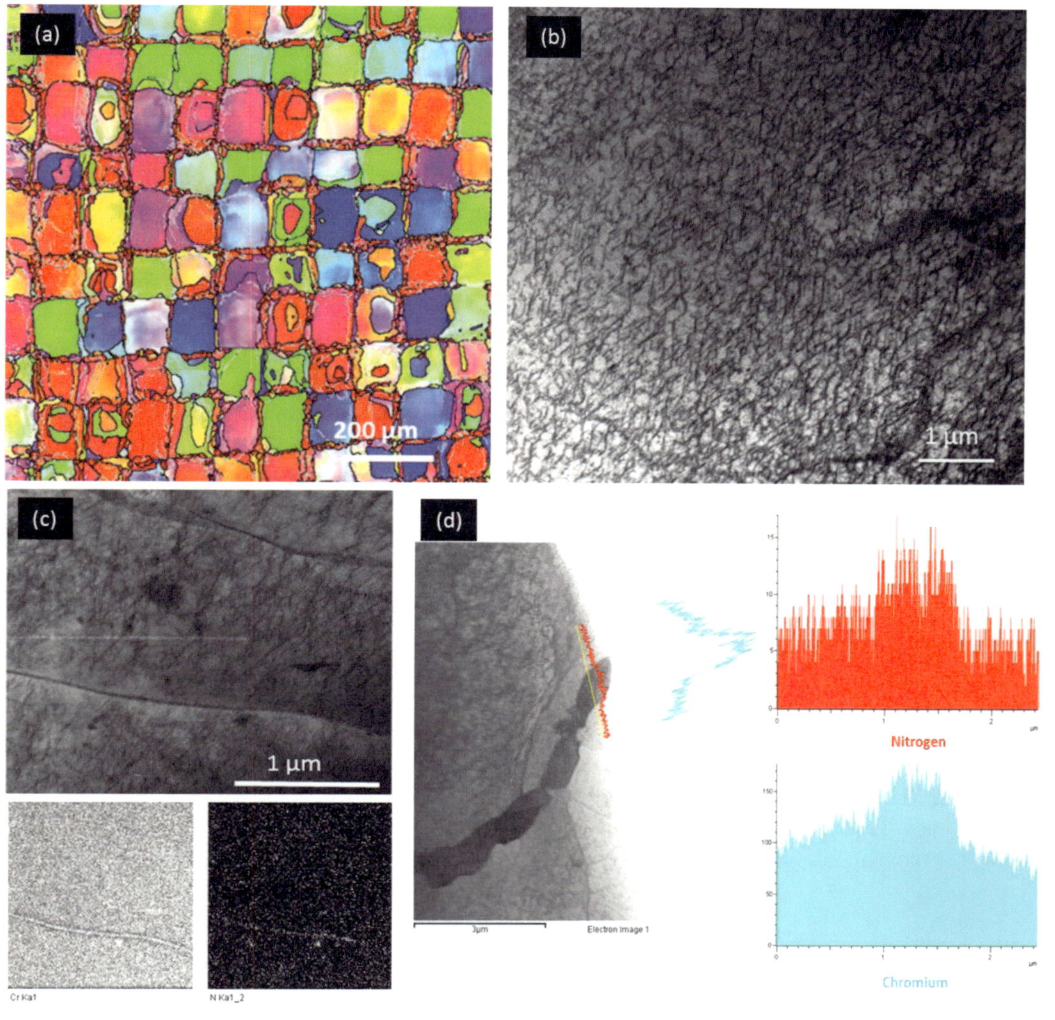

Figure 9. EBSD image of (a) top view of SLM duplex 2507 stainless steel showing a mosaic texture in which each tesserae contains grains, (b) dense 3D network of dislocations forming loops approximately 200 nm, (c) TEM mapping showing nitrogen enriched zones, (d) TEM line scanning revealing nitrides (copyrights obtained; DOI: 10.1016/j.msea.2016.04.027).

SLM fabricated martensitic stainless steel, grade 420, demonstrated hard material with microhardness of 650 Hv and excellent mechanical strength (1800 MPa tensile strength and 800 MPa yield strength) with low toughness i.e., 4–5% of elongation. The strength and microhardness of 420 grade processed by SLM exceeds that of tools steels reported [32, 33]. The superior mechanical strength can be explained by presence of fine cellular structure with submicron martensitic needles. The size of the martensite needles in SLM fabricated martensitic stainless steel, grade 420 is finer than martensitic needles formed by quench and partitioning reported in literature [34, 35]. The colonies of fine cells and submicron martensitic needles can effectively promote the mechanical strength of the SLM 420 material. **Table 3** compares the mechanical properties of SLM 420 with conventional processed AISI 420 stainless steel.

3.2. Mechanical performance and microstructural studies at intermediate temperatures (250–800°C)

The structure materials for in-vessel components of the international thermonuclear experimental reactor (ITER) will require the material to withstand temperature of the working environment (estimated around 200–250°C). 316L stainless steel has been selected as one candidate due to its combination of good mechanical properties, excellent corrosion resistance and good machinability [36, 37]. Challenging issue of the ITER in-vessel components is their geometrical complexity containing inner structure details, like piping systems. These components are difficult to build, costly and sometimes even impossible to build by any conventional manufacturing process. 316L manufactured by SLM shows to be promising for building complex components with enhanced properties. Zhong et al. [38] studied the mechanical properties of SLM 316L at 250°C and made a comparison with the required properties suitable for ITER. **Table 4** shows the mechanical properties of SLM 316L, HIP 316L and the properties required for ITER. As seen, the mechanical properties of the SLM 316L at room temperature and 250°C are higher than the strength requirement for an ITER component. This study successfully confirmed that the SLM was a suitable technique for building complex components with enhanced properties required for ITER application. Well-arranged cellular subgrains strengthen SLM 316L by dislocation pinning effect following Hall-Petch rule and give rise to higher strength and lower ductility than EBM 316L. However larger columnar grains and irregular subgrains in EBM 316L results in higher ductility and higher toughness than SLM 316L [38].

Industrial demands for stainless steel components and parts to be used at higher service temperatures have attracted attentions towards using stainless steels for high temperature applications. Thus, today, steels such as martensitic grade 403 steel is used in gas turbines at 550–600°C [39] and A286 austenitic stainless steel is used in jet engines at 700°C [40]. The need

Type of material	Tensile strength (Mpa)	Yield strength (Mpa)	Elongation (%)	Microhardness (HV)
SLM 420	1800 MPa	800 Mpa	5%	650 HV
Annealed cold drawn AIS 420 [32]	800 MPa	700 Mpa	6-7%	-----

Table 3. Table comparing SLM 420 stainless steel with the mechanical properties of conventionally cold drawn 420 stainless steel.

316L type	Yield Strength (MPa)	Tensile strength (MPa)	Total Elongation (%)
SLM at RT	487 ± 3	594 ± 4	49 ± 4
SLM at 250 °C	376 ± 3	461 ± 3	31 ± 4
HIP at RT/200 °C	220/160	570/450	54/42
Criteria RT/ET	220/135	525/415	45/-
EBM at RT	253 ± 3	509 ± 5	59
EBM at 250 °C	152 ± 3	386 ± 3	46 ± 3

Table 4. Mechanical properties of SLM 316L and EBM 316L at room and 250°C temperature and properties criteria for ITER application.

for a low cost, easy to machine and widely used stainless steel such as 316L has been realized. The 316L is typically utilized at service temperatures up to 500°C [41, 42] whereas above 500°C, the steel undergoes composition and microstructure changes such as precipitation of unwanted phases and grain coarsening that degrades its performance.

Figure 10a shows the mechanical performance of SLM 316L at 800°C. The material withstands tensile stress of 400 MPa before failure compared to 180 MPa for conventional 316L and 304 stainless steels. The tensile test curve is divided into three regions. In the first region the stress reaches a plateau of 100 MPa while simultaneously stretching and elongating to 10%. The reason for this the structural hierarchy presented in the steel microstructure after SLM which remains despite being heated to 800°C. The hierarchy can be verified from the EBSD figures where the melt pools with width of 100 μm are seen and inside the melt pools array of elongated grains are observed which themselves contain smaller subgrains (**Figure 10b** and **c**). In the second region, strain strengthening effect occurs, i.e. after the yield point elongation deformation proceeds uniformly. These phenomena are due to the fine micron sized

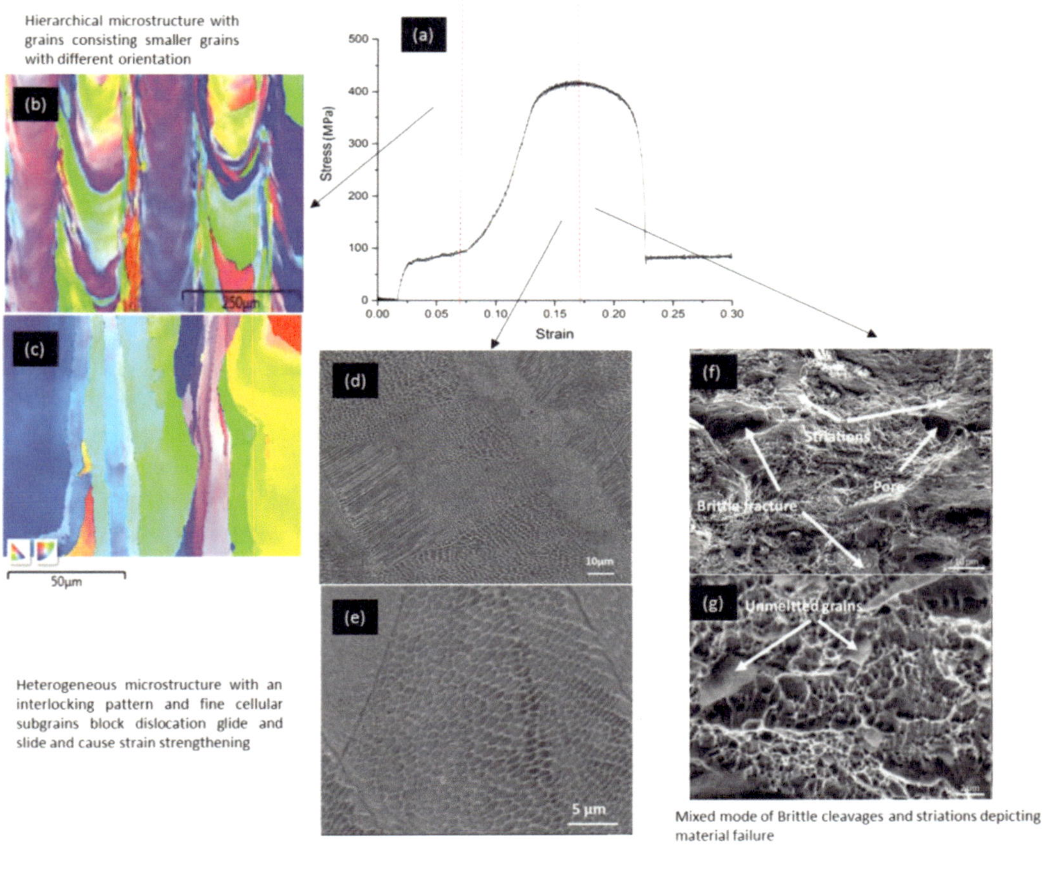

Figure 10. (a) Tensile test curve of SLM 316 L obtained at 800°C and divided into three regions, (b and c) in the first region the hierarchy verified from the EBSD figures cause strengthening, (d and e) fine micron sized subgrains interacting with dislocation movement, (f and g) fracture analysis showing a dual ductile-brittle fracture mode with striations, cup and cone dimples and brittle cleavages (copyrights obtained; DOI: 10.1016/j.matdes.2017.08.072).

subgrains (see **Figure 10d** and **e**) inhibiting dislocation movement. These phenomena are typical in materials with low stacking fault energies such as austenitic stainless steels. In the third region threshold value of the material is reached after 400 MPa stress and the materials fails in a dual ductile-brittle mode. **Figure 10f** and **g**) shows striations, cup and cone dimples and brittle cleavages confirming a dual failure mode.

3.3. Mechanical performance and microstructural studies of SLM stainless steels at high temperatures (>800°C)

The microstructure of a polycrystalline material is a key parameter in determining a wide range of its properties including mechanical strength, toughness, electrical conductivity and magnetic susceptibility [43, 44]. One important aspect of microstructure is the size of the grains and its influence on the properties. Thus, to design the grain size and restrict grain growth, microstructural engineering has become of fundamental importance [45]. High temperature applications are critically influenced by grain growth, precipitation and phase changes. Biggest issue which is very common and probable in materials exposed at high temperatures is the abnormal grain growth, where one or more grains will grow abnormally in the microstructure. The presence of abnormally large grains is detrimental to the mechanical properties of polycrystalline structures [46]. Mechanical properties of SLM 316L at 1100 and 1200°C is listed in **Table 5** and compared to conventional 316L materials at the same temperature [47].

The Superior, 10 times higher, mechanical strength of 300 and 150 MPa of 316L SLM (for temperatures 1100 and 1200°C respectively) compared to 35–40 and 20 MPa, points out that the rapid grain growth and loss of strength has been avoided. The growth mechanism involved in this case is "subgrain controlled mechanism" shown by the schematic in **Figure 11a**. In step one, the cellular subgrains inside and confined by the macro grains (shown by the hexagon) are seen. During heating at temperatures up to 1100°C (step 2) the subgrain boundaries merge with adjacent subgrains and form larger subgrains while still being confined by the larger macro grains. In step three, upon heating to temperatures ≥1200°C the cellular subgrains have disappeared and new subgrains with irregular shapes have formed which are bigger than 1 μm and are still confined inside the macro grains. This continues until at very high temperatures (>1200°C) the subgrains have grown into the size of the macro grain and from this point on the macro grains will start to grow and abnormal grain growth can possibly occur. SEM microscopy observation shows the clear tendency of subgrains merging into bigger

Type	Grain size (μm)	Tensile strength (MPa)	Elongation (%)
SLM-1100°C	5-16 μm	300 MPa	15-18%
SLM-1200°C	32 μm	150 MPa	20%
Casted-1100°C	50 μm	35-40 MPa	40%
Casted-1200°C	80 μm	20 MPa	45%

Table 5. Mechanical properties of SLM 316L at 1100 and 1200°C and comparison to conventional 316L materials at the same temperature.

subgrains as seen in **Figure 11b** and **c**. The white dashed areas show how the 1 μm-sized cells merge and become a bigger subgrain (approx. 2–5 μm) and the black line marks the boundary of the initial macro sized grain (approx. 10–20 μm) containing the 1 μm subgrains. By applying higher temperature (1200°C), it appears that the subgrains have disappeared and the irregular macro sized grains have remained (see **Figure 11d** and **e**).

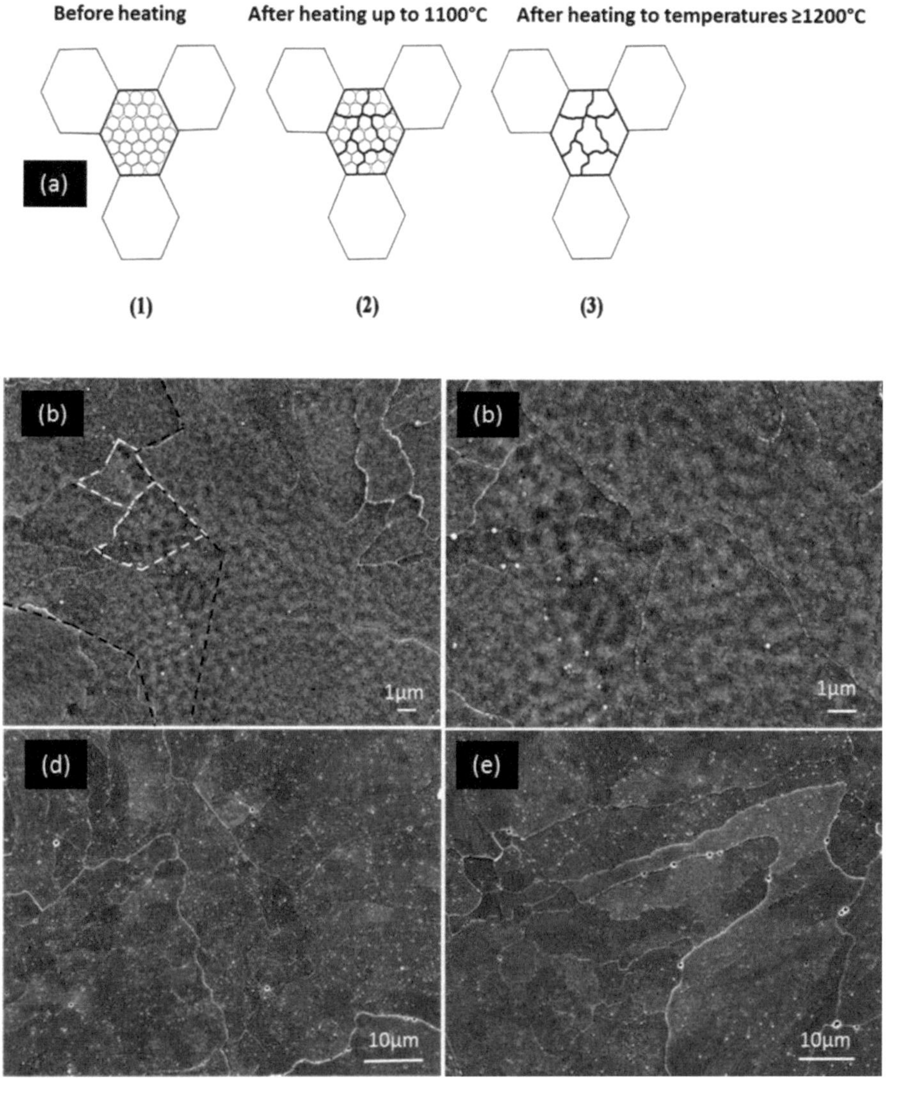

Figure 11. (a) Schematic of the grain growth mechanism in SLM 316L showing three stages before heating, after heating to temperatures up to 1000°C and after heating to temperatures higher than 1200°C, (b and c) SEM microscopy observation shows the clear tendency of subgrains merging into bigger subgrains upon heating to temperature as high 1100°C. The white dashed areas show how the 1 μm-sized cells merge and become a bigger subgrain (approx. 2–5 μm) and the black line marks the boundary of the initial macro sized grain (approx. 10–20 μm) containing the 1 μm subgrains, (d and e) the subgrains have disappeared and the irregular macro sized grains have remained (copyrights obtained; DOI: 10.1098/rsos.172394).

High temperature mechanical properties of SLM 2507 duplex was assessed at 1200°C. The tensile test curve is seen in **Figure 12a**. Comparison with conventional duplex material made in **Table 6** shows significantly higher strength of SLM 2507 material with tensile strength of 200 MPa. The high strength is due to the fine microstructure consisting of columnar grains of austenite (100 μm in length and 2–4 μm in width) and ferrite grains in between them (**Figure 12b**). The interlocking network strengthens and toughens the material at the same time, showing 30% elongation before failure. The ductile failure mode is confirmed by the cup and cone dimples seen in **Figure 12c** and **d**.

In another study the SLM duplex 2507 was first heat treated to 1200°C in a tube furnace with cooling rate 300°C/hr thereafter mechanical properties of the material was measured at 1200°C. The mechanical strength is around 400 MPa (**Figure 13a**) which is two times higher than one reported in **Figure 12**. Intermetallic phases such as sigma precipitate due to enrichment of Mo and Cr in the matrix of stainless steel (**Figure 13b** and **c**). Due to non-equilibrium nature of SLM 2507 and the high activation energy of Mo and Cr for diffusion, upon heating the material to high temperatures such as 1200°C the diffusion barrier of Mo is met and activation energy for the diffusion is provided thus phase changes and sigma precipitation can occur upon cooling.

Fast cooling rates during heat treatment can influence the formation of intermetallic sigma phase. Studies carried out [48] and shown by TTT diagrams indicate cooling rates (higher

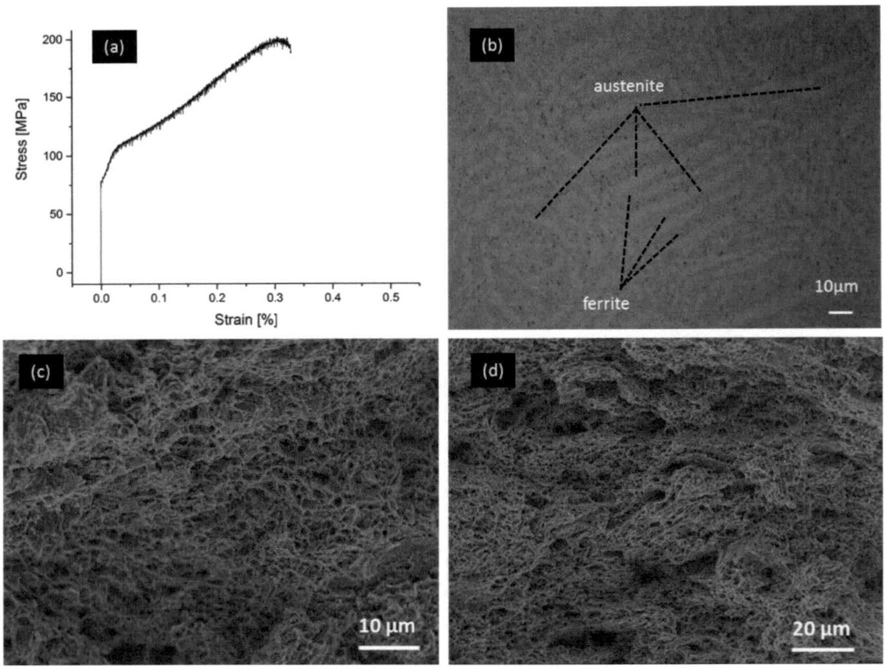

Figure 12. (a) Tensile test curve of SLM 2507 obtained at 1200°C, (b) fine microstructure consisting of columnar grains of austenite (100 μm in length and 2–4 μm in width) and ferrite grains in between them, (c) fracture analysis showing ductile failure mode confirmed by the cup and cone dimples (figures have been used in paper submitted to Journal of Alloys and Compounds on 20/06/2018.).

http://dx.doi.org/10.5772/intechopen.80468

Tensile test at 1200 °C	Tensile strength (MPa)	Elongation (%)
SLM duplex 2507 (without sigma precipitates)	200 MPa	30%
Conventional duplex	19 MPa	----

Table 6. Mechanical properties of duplex stainless steel obtained at 1200°C.

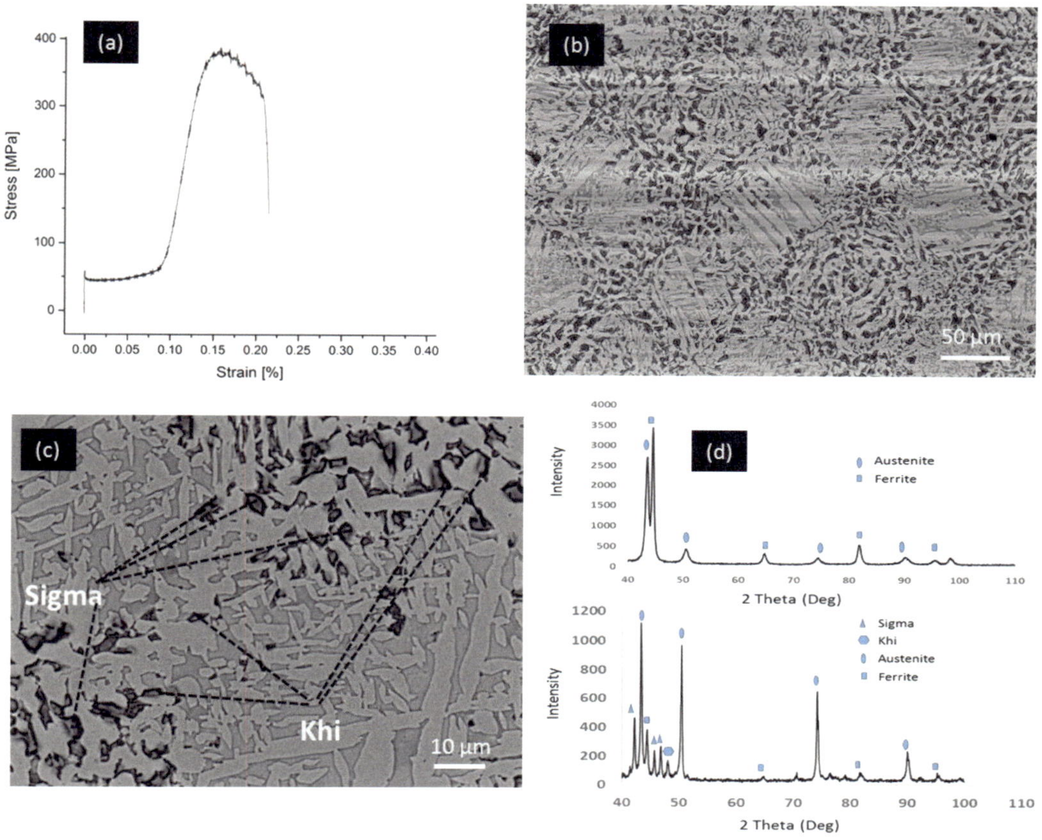

Figure 13. (a) Tensile test curve of heat treated SLM 2507 obtained at 1200°C, (b and c) intermetallic sigma and Khi precipitates, (d) XRD pattern of SLM 2507 with intermetallic and without intermetallic formation (figures have been used in paper submitted to Journal of Alloys and Compounds on 20/06/2018).

than 250°C/hr) can completely avoid sigma precipitation. The test instrument for the high temperature mechanical testing is Gleeble^{3800}. The cooling rate in vacuum in the Gleeble chamber is approximately 200°C/s which is much higher than 250°C/hr and thus sigma can totally be avoided. Therefore, the material heat treated at slow cooling rates precipitates sigma phase, which are strong and brittle phase precipitates (see **Figure 13b** and **c**). Formation of sigma phase strengthens the material and causes higher strength as compared to material with no sigma phase (see XRD patterns in **Figure 13d**).

4. Conclusions

Tailoring microstructure, tuning the chemistry, adjusting phase composition and introducing dense 3D network of dislocations through fast cooling processes such as selective laser melting can tailor and develop stainless steels with high performance for extreme conditions, such as elevated temperatures. The very fine cellular sub-grains and Mo enrichment at the cell boundaries can enable the material to withstand more stress and higher temperatures. The non-equilibrium phase composition and phase transformations obtained after laser melting can strengthen the material via formation of coarser martensitic needles (in 420 stainless steel) and intermetallic phases (in 2507 duplex stainless steel) upon post heat treatment.

Acknowledgements

The authors would like to thank the funding support from the Carl Tryggers Foundation (grant no. CTS 16:6) and the Swedish Foundation for Strategic Research (SSF) for an infrastructure fellowship grant (RIF14-0083).

Conflict of interest

The authors declare that they have no conflict of interests.

Notes/Thanks/Other declarations

The authors would like to thank Emil Hedin, Valeri Ivanov and Daniel Zapata for sample preparations and helping out with the characterizations.

Author details

Kamran Saeidi* and Farid Akhtar

*Address all correspondence to: kamran.saeidi@ltu.se

Department of Mathematics and Engineering Sciences, Division of Engineering Materials, Luleå University of Technology, Luleå, Sweden

References

[1] Gibson I, Rosen DW, Stucker B. Additive Manufacturing Technologies. NewYork, USA: Springer; 2010

[2] Feygin M, Pak SS. Laminated Object Manufacturing Apparatus and Method. Google Patents; 1999

[3] Zein I, Hutmacher DW, Tan KC, Teoh SH. Fused deposition modeling of novel scaffold architectures for tissue engineering applications. Biomaterials. 2002;**23**(4):1169-1185

[4] Jacobs PF. Rapid Prototyping & Manufacturing: Fundamentals of Stereolithography. Society of Manufacturing Engineers. NewYork, USA: McGraw-Hill; 1992

[5] Ibrahim D, Tiago LB, Heitz C, De Oliveira MG, De Oliveira HW. Dimensional error of selective laser sintering, three-dimensional printing and PolyJet™ models in the reproduction of mandibular anatomy. Journal of Cranio-Maxillofacial Surgery. 2009; **37**(3):167-173

[6] Stotko CM. Laser sintering: Layer by layer. Nature Photonics. 2009;**3**(5):265-266

[7] Kruth J-P, Froyen L, Van Vaerenbergh J, Mercelis P, Rombouts M, Lauwers B. Selective laser melting of iron-based powder. Journal of Materials Processing Technology. 2004; **149**(1):616-622

[8] Gu D. Laser Additive Manufacturing of High-Performance Materials. Heidelberg, Germany: Springer; 2015

[9] Cormier D, Harrysson O, West H. Characterization of H13 steel produced via electron beam melting. Rapid Prototyping Journal. 2004;**10**(1):35-41

[10] Heinl P, Muller L, Korner C, Singer RF, Muller FA. Cellular Ti–6Al–4V structures with interconnected macro porosity for bone implants fabricated by selective electron beam melting. Acta Biomaterialia. 2008;**4**(5):1536-1544

[11] Harrysson OL, Kansyzoglu O, Marcellin-little DJ, Cormier DR, West HA. Direct metal fabrication of titanium implants with tailored materials and mechanical properties using electron beam melting technology. Materials Science and Engineering C. 2008; **28**(3):366-373

[12] Atwood C. Laser Engineered Net Shaping (LENS (TM)): A Tool for Direct Fabrication of Metal Parts. Albuquerque, NM/Livermore, CA: Sandia National Laboratories; 1998

[13] Utela B, Storti D, Anderson R, Ganter M. A review of process development steps for new material systems in three dimensional printing (3DP). Journal of Manufacturing Processes. 2008;**10**(2):96-104

[14] Saeidi K. Stainless steels fabricated by laser melting: Scaled down structural hierarchies and microstructural heterogeneities [thesis]. Stockholm University, Department of Materials and Environmental Chemistry; 2016

[15] Beng CH. Selective laser sintering of a stainless steel powder [thesis]. University of Leeds. Leeds, England: School of Mechanical Engineering; 2003

[16] Miranda G, Faria S, Bartolomeu F, Pinto E, Madeira S, Mateus A. Predictive models for physical and mechaical properties of 316L stainless steel produced by selective laser melting. Materials Science Engineering A. 2016;**657**:43-56

[17] Sun Z, Tan X, Tor SB, Yeong WY. Selective laser melting of stainless steel 316L with low porosity and high build rates. Materials and Design. 2016;**104**:197-204

[18] Saeidi K, Gao X, Zhong Y, Shen J. Hardened austenite steel with columnar sub-grain structure formed by laser melting. Journal of Materials Science Engineering A. 2015; **625**:221-229

[19] Wang H, Sweikart MA, Turner JA. Stainless steel as bipolar plate material for polymer electrolyte membrane fuel cells. Journal of Power Sources. 2003;**115**:243-251

[20] Dewidar MM, Khalil KA, Lim J. Processing and mechanical properties of porous 316L stainless steel for biomedical applications. Transactions of Nonferrous Metals Society of China. 2007;**17**:468-473

[21] Xu FL, Duan JZ, Lin CG, Hou B. Influence of marine aerobic biofilms on corrosion of 316L stainless steel. Journal of Iron and Steel Research, International. 2015;**22**:715-720

[22] Kastelic J, Galeski A, Baer E. The multicomposite structure of tendon. Connective Tissue Research. 1978;**6**:11-23

[23] Weiss PS. Hierarchical assembly. ACS Nano. 2008;**2**:1085-1087

[24] National Research Council and others. Hierarchical Structures in Biology as a Guide for New Materials Technology. Washington, DC: National Academy Press; 1994. p. 464

[25] Vogel F, Wanderka N, Balogh Z, Ibrahim M, Stender P, Schmitz G, Banhart J. Mapping the evolution of hierarchical microstructures in a Ni-based super alloy. Nature Communications. 2013;**4**:1-7

[26] Song JE. Hierarchical multisacle modeling of Ni-base super alloys [Master of Science thesis]. Georgia Institute of Technology. Atlanta, USA: School of Mechanical Engineering; 2010

[27] Kurella AK. Laser induced hierarchical coatings on titanium alloy [Doctor of Philosophy thesis]. Knoxville: University of Tennessee; 2009

[28] Liu G, Dingdong F, Li J, Chen K, He G, Yang Z, Guo S. Combustion synthesis of W–Cr alloys with hierarchical microstructure and improved hardness. Materials Letters. 2016; **166**:43-45

[29] Morris Wang Y, Voisin T, McKeown JT, Ye J, Calta NP, Li Z, et al. Additively manufactured hierarchical stainless steels with high strength and ductility, Nature Materials. 2018;**17**:63-71

[30] Seritova M. Handbook of Stainless Steel. Espoo, Finland: Autokumpu; 2103

[31] Yan FK, Liu GZ, Tao NR, Lu K. Strength and ductility of 316L austenitic stainless steel strengthened by nano-scale twin bundles. Acta Materialia. 2012;**60**:1059-1071

[32] Chen H, Gu D, Dai D, Ma C, Xia M. Microstructure and composition homogeneity, tensile property, and underlying thermal physical mechanism of selective laser melting tool steel parts. Materials Science and Engineering: A. 2017;**682**:279-289

[33] Brnic J, Turkalj G, Canadija M, Lanc D, Krscanski S. Martensitic stainless steel AISI 420—Mechanical properties creep and fracture toughness. Mechanics of Time-Dependent Materials. 2011;**15**:341-352

[34] Kitahara H, Ueji R, Tsuji N, Minamino Y. Crystallographic features of lath martensite in low-carbon steel. Acta Materialia. 2006;**54**:1279-1288

[35] Lu SY, Yao KF, Chen YB, Wang MH, Chen N, Ge XY. Effect of quenching and partitioning on the microstructure evolution and electrochemical properties of a martensitic stainless steel. Corrosion Science. 2016;**103**:95-104

[36] Kalinin G, Barabash V, Cardella A, Dietz J, Ioki K, Matera R. Assessment and selection of materials for ITER in-vessel components. Journal of Nuclear Materials. 2000;**283**:10-19

[37] Tavassoli AA. Assessment of austenitic stainless steels. Fusion Engineering and Design. 1995;**29**:371-390

[38] Zhong Y, Rännar LK, Wikman S, Koptyug A, Liu L, Cui D, Shen Z. Additive manufacturing of ITER first wall panel parts by two approaches: Selective laser melting and electron beam melting. Fusion Engineering and Design. 2017;**116**:24-33

[39] Muktinutalapati NR. Materials for gas turbines—An overview. In: Benini E, editor. Advances in Gas Turbine Technology. InTech; 2011. pp. 293-314

[40] De CH, Luppo MI, Raffaeli H, Di Gaetano J, Gribaudo LM, Ovejero-García J. Creep behavior of an A286 type stainless steel. Materials Characterization. 2005;**55**:97-105

[41] Hoke JH. Mechanical properties of stainless steels at elevated temperatures. Handbook of Stainless Steels. NewYork, USA: MacGraw-Hill; 1977

[42] Davies CM. Predicting creep crack initiation in austenitic and ferritic steels using the creep toughness parameter and time-dependent failure assessment diagram. Fatigue and Fracture of Engineering Materials and Structures. 2009;**32**(10):820-836

[43] Li JCM. Microstructure and Properties of Materials. Vol. 1. Singapore: World Scientific Publishing; 1996. ISBN: 978-981-02-2403-5

[44] Penn SJ, Alford MN, Templeton A, Wang X, Xu M, Reece M, Scharp K. Effect of porosity and grain size on the microwave dielectric properties of sintered alumina. Journal of American Ceramic Society. 1997;**80**:1885-1888

[45] Davidson KP, Singamneni S. Magnetic characterization of selective laser melted SAF 2507 duplex stainless steel. Journal of Metals. 2017;**69**:569-574

[46] Dennis J, Bate PS, Humphreys JF. Abnormal grain growth in metals. Material Science Forum. 2007;**558-559**:717-722

[47] Saeidi K, Akhtar F. Sub-grain controlled grain growth in laser melted 316L promoting strength at high temperatures. Royal Society Open Science. 2018;**5**:172394

[48] Padilha AF, Aguiar DJM, Plaut RL. Duplex stainless steels: A dozen of significant phase transformations. Defect Diffusion Forum. 2012;**322**:163-174

Chapter 8

Entropic Alloys for Cryogenic Applications

Rui Xuan Li and Yong Zhang

Additional information is available at the end of the chapter

http://dx.doi.org/10.5772/intechopen.82351

Abstract

The entropic alloys can be categorized into four types of alloys, e.g., high-entropy alloys, medium-entropy alloys, low-entropy alloys, and pure metals. The high-entropy alloys are a new kind of materials where the mixing entropy plays an important role in the phase formation. Because of the unique structures, the entropic alloys exhibit many outstanding properties, which even break the performance limits of traditional materials, including the excellent low-temperature properties. The mechanical properties of the entropic alloys serving at low temperature are mainly introduced in this chapter, including strength, plasticity, fracture behaviors, and impact resistance, and the reasons for these behaviors reported in recent years are also summed up.

Keywords: entropic alloys, low-temperature brittleness, nano-twins, phase transformation, cryogenic applications

1. Introduction

Materials usually exist in the lowest energy state. It has been clarified in the thermodynamic equation G = H − TS (where G is Gibbs free energy, H is the enthalpy and S is the entropy of the system) that, to reduce the total energy of the system, we can increase T or S, or decrease H, corresponding to the superalloys, high entropy alloys and intermetallics respectively. As a result, the entropy plays an important role in the alloy system, which can be used to define the alloys.

The entropic alloys can be a new categorizing of the alloys filed, which mainly concern the configuration of the components in the alloy systems. The configurational entropy of the alloy can use the Boltzmann equation:

$$\Delta S = kLnW \quad (1)$$

Here, k is Boltzmann constant and W is the number of real microstates corresponding to the macrostate.

The entropic alloys can be categorized into four types of alloys as in **Figure 1**:

1. High-entropy alloys: The configurational entropy ΔS is above 1.61R; here R is gas constant.
2. Medium-entropy alloys: 0.69R < ΔS < 1.61R.
3. Low-entropy alloys: ΔS < 0.69R.
4. Ultrapure materials: ΔS = 0.

The high-entropy alloys are a new type of multicomponent materials, which were first proposed by Professor J. W. Yeh of National Tsing Hua University in 2004 [1]. Traditional alloys are usually based on one or two principal elements (usually having a principal content greater than 50%), with the alloy design range locating on the edge of the phase diagram. However, it is broken by the appearance of entropic alloys, and the alloy design range extends to the center of the phase diagram successfully. The entropic alloys usually consist of five or more components, and the atomic percentage of each component is 5–35% with no difference between the solute and solvent [2, 3]. According to the traditional physical metallurgy principle and the Gibbs phase law [4], the more the number of components containing in the alloy, the more likely it is to have a variety of solid solutions and intermetallic compounds. However, in the entropic alloys, the large mixing entropy plays a dominant role to expand the dissolution range of the intermetallic compound and the terminal solid solution, and instead a relatively simple phase structures of face-centered cubic (FCC), body-centered cubic (BCC), or close-packed hexagonal (CPH) are obtained [5–7]. As reported in Yeh's work, when the number of components of the Cu-Ni-Al-Co-Cr-Fe-Si alloy system is increased to 7, only FCC and BCC

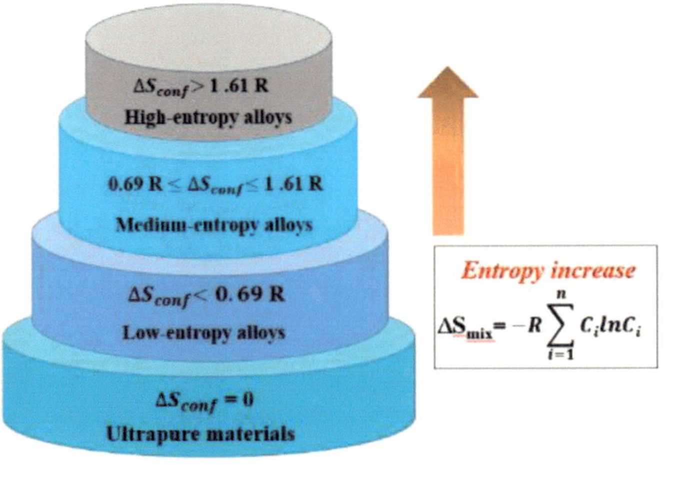

Figure 1. The classification of entropic materials.

simple solid solution structures are still formed [8]. Based on the important role the mixing entropy played in the alloy, it is widely accepted that a molten alloy system where the effect of entropy plays the dominant role ($\Omega \geq 1.1$) and which can form the simple solid solution can be called the entropic alloys [9].

There are a large number of metallic and nonmetallic elements that can be involved in the entropic alloys, including some main group elements in the IIA, IIIA, and IVA group (such as B, Al, Si, Mg, etc.), all the subgroup elements in the fourth period (such as Ti, Cr, Fe, Mn, Co, Ni etc.), and some subgroup elements in the fifth period (such as Zr, Mo, Nb, etc.). Furthermore, there are four core effects in the entropic alloys, including high-entropy effect in thermodynamics, late diffusion effect in dynamics, lattice distortion effect in structure, and cocktail effect in performance. Such a variety of elemental types and the four core effects result in a variety of alloy systems with different properties, which even break the limits of traditional materials, such as high strength [10], high hardness [11], high radiation resistance [12], high temperature softening resistance [13], high wear resistance [14], fatigue resistance [15], etc.

With the rapid development of technology, low-temperature materials are increasingly used in aerospaces, superconducting fields, and civil industries. These kinds of materials can be used in a wide temperature range from below room temperature to absolute zero. It is well known to us that the strength tends to increase while the plasticity and toughness significantly decrease when the conventional materials serve at the low temperature, which in turn results in the cold-brittle fracture. This cold-brittle fracture is similar to the brittle fracture at normal temperature. There is no obvious plastic deformation before fracture, and it suddenly occurs. The crack usually originates from defects or stress concentrations in the material structures and rapidly expands. The type becomes transgranular cleavage, and the fracture feature changes from fibrous to crystalline. This cold brittleness is related to the lattice type of the material. Generally, the BCC and HCP structures are cold-brittle materials, while the FCC metals are non-cold-brittle materials. For the entropic alloys with many outstanding properties and with great application potential in many fields, the researches on low-temperature performance are also in full swing. Therefore, the research status and application prospects of low-temperature properties of the entropic alloys are summarized in this paper, mainly aimed at mechanical properties including strength and fracture behavior.

2. The research status of low-temperature mechanical properties

2.1. The strength and plasticity at low temperature

The most important requirement for materials that serves at low temperatures is to have sufficient strength and stiffness, and in the view of minimizing the total equipment weight, it is often desirable to have the yield strength as high as possible.

There have been a large number of reports about the excellent room temperature tensile properties of the entropic alloys. The crystal structures of the entropic alloys have a significant influence in the tensile properties. Alloys with BCC structures usually have high strength and low plasticity, while alloys with FCC structures are usually accompanied by the large plastic toughness and low strength. However, as the entropic alloys with the BCC structures usually turn to extreme brittleness at the low temperature, researches are mainly aiming at BCC entropic alloys. The earliest research on the low-temperature tensile properties should belong to the E.P. George group of Oak Ridge National Laboratory [16, 17]. Starting from 2013, they systematically studied the low-temperature tensile properties of the FCC entropic alloys and found that the FCC CoCrFeNiMn alloys are characterized by stronger with lower temperature in a wide temperature range. **Figure 2** shows the strength-temperature curves of CoCrFeNiMn and CoCrFeNi alloys at two different strain rates. It can be seen that the yield strengths of these two kinds of alloys have higher temperature dependence and lower strain-rate dependence, and the temperature dependence becomes stronger at lower temperatures (−200–0°C) and higher temperatures (600–1000°C). As the temperature decreases, the yield strength of both alloys increases rapidly, while the elongation does not decrease as the conventional alloy, as shown in **Figure 3**. This abnormal increase of both strength and ductility contributes to the deformation mechanism transforming from dislocation slip to nano-twins, which will be explained later in Section 3.

2.2. The fracture behavior at low temperature

Fracture toughness characterizes the ability of an alloy to resist crack propagation and is a quantitative indicator of the toughness. When the crack size is constant, the larger the fracture toughness value of an alloy, the greater the critical stress required for the crack instability and expansion. When the external force is given, if the fracture toughness value is higher, the critical size of the crack to reach the instability expansion is larger. The fracture toughness can be used as a supplement to the materials which are designed according to yield strength

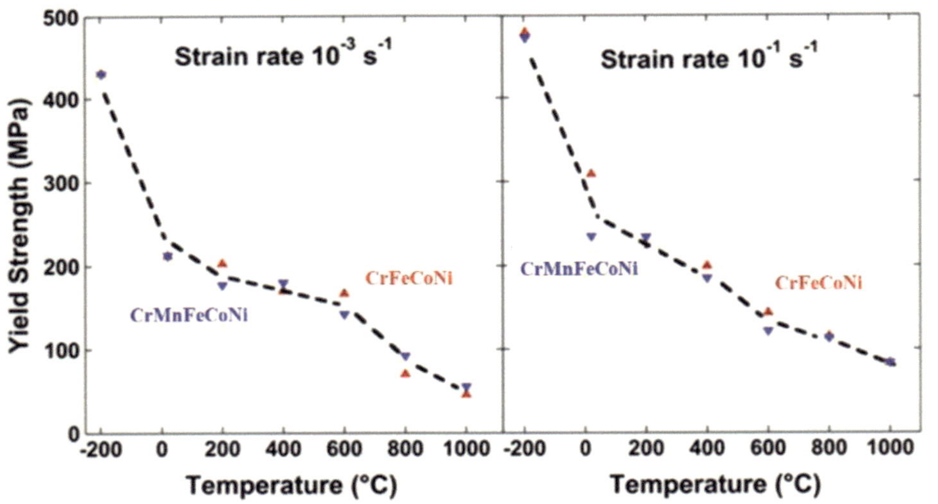

Figure 2. The strength-temperature curves of CoCrFeNiMn and CoCrFeNi alloys at two different strain rates [16].

http://dx.doi.org/10.5772/intechopen.82351

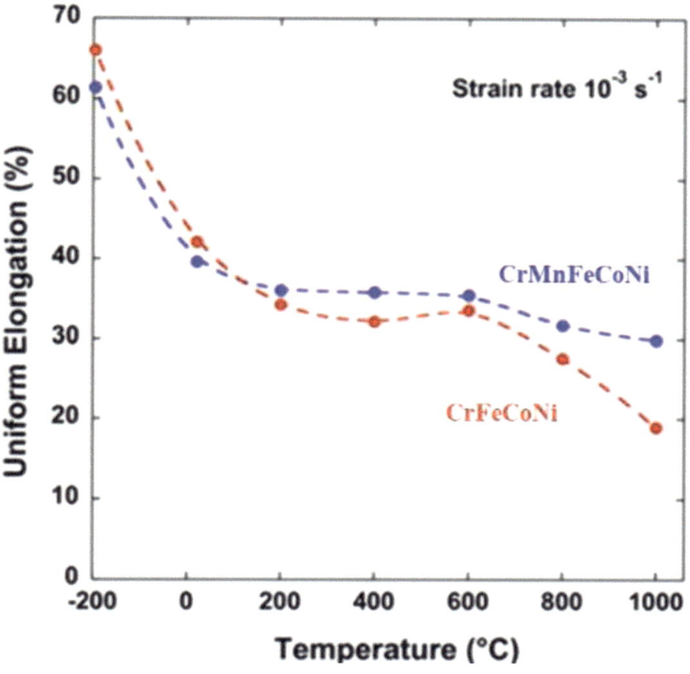

Figure 3. Temperature dependence of the tensile ductilities of CrMnFeCoNi and CrFeCoNi tested at an engineering strain rate of 10^{-3} s^{-1} [16].

criteria. For example, the fracture toughness values usually decrease with the increasing yield strength. Therefore, although the structure of the high-strength alloys has been subjected to a very safe design according to the yield strength criterion, the unstable fracture and fatal damage may occur simply because of an inner crack. Furthermore, since the yield strength of the material increases with decreasing temperature, the low-temperature fracture toughness is not considered in the design process, and the same result may also be caused. Therefore, the characterization of the fracture toughness of entropic alloys is also important.

Professor R.O. Ritchie from the Lawrence Berkeley National Laboratory found the extremely high fracture toughness of FCC entropic alloys [18]. The related research work was published on Science, which arouse widespread concerns of the entropic alloys. It is reported that the CoCrFeMnNi alloys, which are rolled and completely recrystallized, have a tensile strength of 1280 MPa at 77 K, which is 70% higher than that at room temperature. More importantly, its plasticity has been significantly improved, which has fracture strain value greater than 70% at 77 K, 25% higher than that of room temperature, and the work hardening index is as high as 0.4. It also has good fracture toughness. As shown in **Figure 4(a)**, as the temperature increases from 77 to 200 K and then to 293 K, the crack-initiation fracture toughness of CrMnFeCoNi entropic alloys increases slightly from 219 to 221 MPa $m^{1/2}$ and then slightly reduced to 217 MPa $m^{1/2}$. After crack initiation, the fracture toughness of the alloy increases rapidly to 307 MPa $m^{1/2}$ with the growth of a large number of subcritical

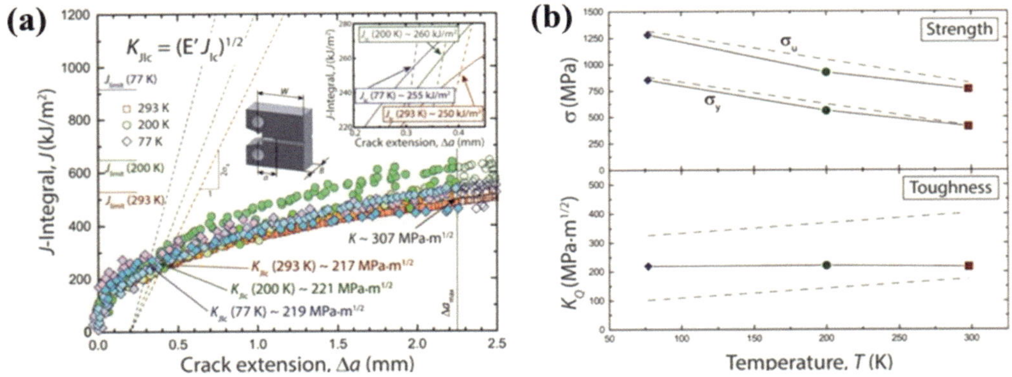

Figure 4. (a) The fracture toughness values change with temperature; (b) the strength and toughness change with temperature [18].

cracks. This high toughness value of entropic alloys can be compared to many high-alloying austenitic stainless steels, such as 304 and 316 L, which has toughness range as K_Q = 175–400 MPa $m^{1/2}$ at room temperature, as well as the best low-temperature steels, which has K_Q = 100–325 MPa $m^{1/2}$ at 77 K. As shown in **Figure 4(b)**, these materials are similar to entropic alloys, and their strengths increase with decreasing temperature. In terms of fracture toughness trends, the toughness of entropic alloys does not change much with temperature, while that of the steels decrease with decreasing temperature.

Subsequent research of this group further found that CoCrNi alloy with equal atomic ratio even has better mechanical properties than CoCrFeMnNi in low-temperature environment, and the fracture toughness at 77 K can reach 273 MPa $m^{1/2}$ [19], which indicates that it is the nature of elements in complex solid solutions rather than their mere numbers that is more important. Indeed, in terms of (valid) crack-initiation and crack-growth toughnesses, the CrCoNi medium-entropy alloy represents one of the toughest materials in any material class ever reported, as manifested in **Figure 5**.

2.3. Impact resistance at low temperature

When the materials are applied at low temperatures, they will be subjected not only to conventional tensile/compression loads but also to the impact loads. Therefore, it is necessary to test the impact toughness by using a pendulum test. The higher the impact toughness is, the more impact energy is absorbed, indicating that the material's ability to withstand impact loads is better.

Utilizing the Charpy impact test standardized in the ASTM standard E-23, Li et al. found that the homogeneous structures of the $Al_{0.1}$CoCrFeNi and $Al_{0.3}$CoCrFeNi entropic alloys after hot forging are obtained, the casting defects are reduced, and the excellent tensile properties and impact resistance are exhibited at low temperatures [20]. The V-notch impact test at 77–298 K shows that the Charpy impact energy of $Al_{0.1}$CoCrFeNi alloy decreases from 420 to 289 J with the decrease of temperature, while it reduces from 413 to 328 J in $Al_{0.3}$CoCrFeNi, which is significantly better than traditional materials such as stainless steels and titanium alloys (as shown in **Figure 6**).

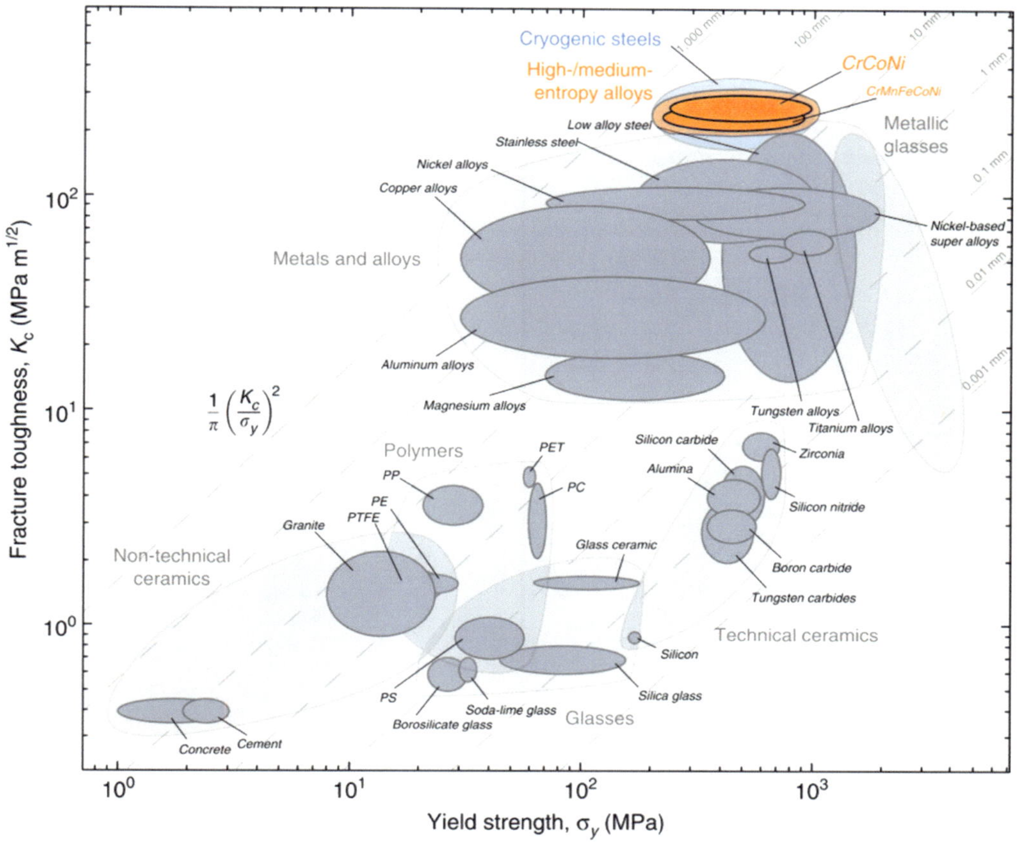

Figure 5. Map of fracture toughness versus yield strength for various classes of materials [19].

However, the Charpy impact test did not exhibit the same increase in strength and plasticity at low temperatures just as the tensile test, which may be owing to the sharp increase in the energy required for dislocation slip under high-speed deformation conditions.

In the impact performance experiment, the type of notch is closely related to the impact toughness. Different types of notch make the stress concentration different. For example, the V-notch reflects the crack propagation, while the U-shaped gap reflects the initiation. At the same time, the state of the material also affects the impact properties. The impact resistance of the Al_xCoCrFeNi after hot forging is reported by Li et al., and the researches on the as-cast Al_xCoCrFeNi (x = 0, 0.1, 0.75, 1, 5) alloys are carried by Xia et al. [21]. As the Al content increases, the alloy changes from a FCC structure (Al_0 and $Al_{0.1}$) to a BCC structure ($Al_{0.75}$ and Al_1), which then bring huge changes to the impact toughness of the alloys. The impact toughness of the as-cast cubic Al_0 and $Al_{0.1}$ alloys at low temperatures is slightly poorer than that of hot-forged alloys [20], but it still has excellent impact resistance, and the impact energy increases with temperature decrease, which have opposite tendency from that in the hot-forged alloys. The impact energy of $Al_{0.1}$ at 77 K is slightly lower than that of Al_0, but it becomes slightly higher as temperature increases.

The impact energy of Al_0 at 77 K is 398 J, which is the best in all alloys. And the average impact energy of the V-notch sample is slightly higher than that of the U-notch sample. It can be seen from the morphology of the sample after the impact tests that Al_0 and $Al_{0.1}$ are destroyed from the notch without broken into two pieces, while the other two alloys were completely impact broken, indicating that their impact resistance are absolutely poor (**Figure 7(a)** and **(b)**).

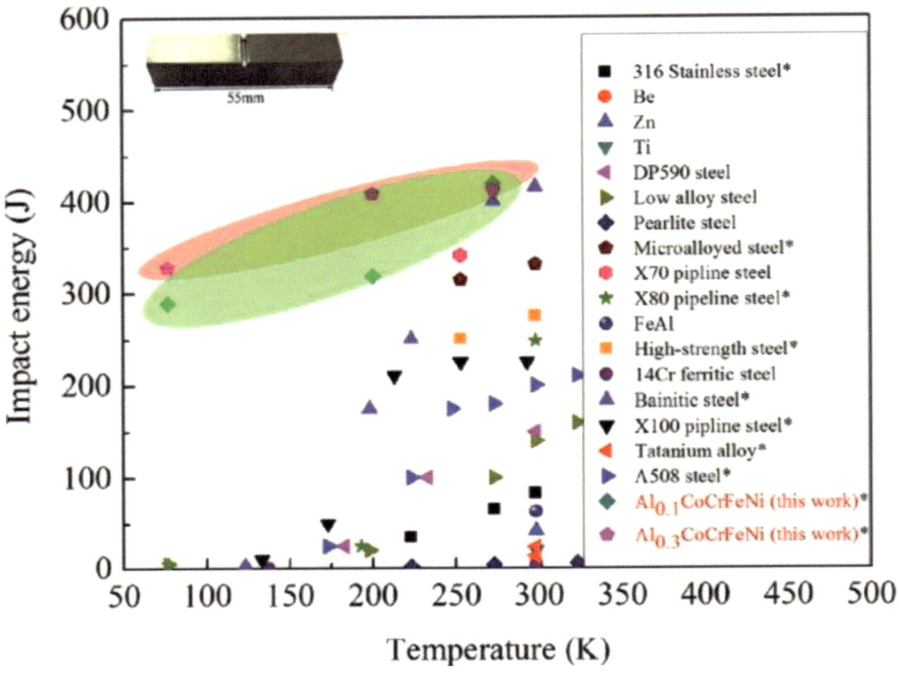

Figure 6. Summary of the Charpy impact energy of materials at the different temperatures [20].

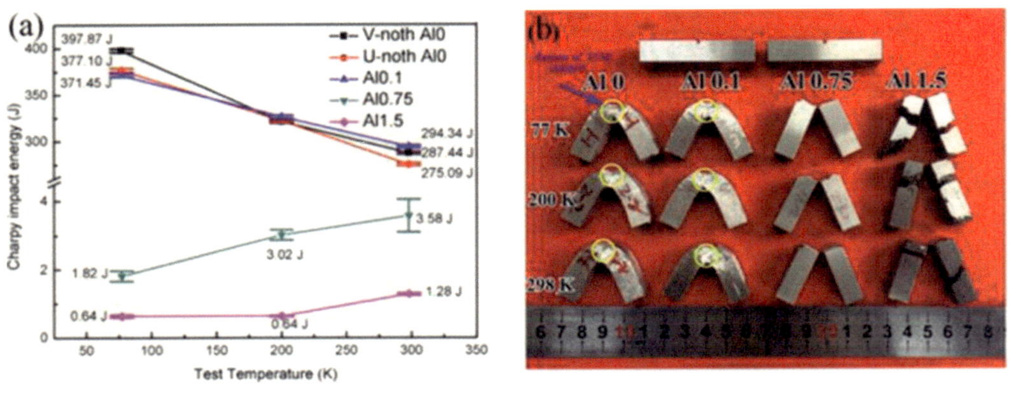

Figure 7. (a) The impact energy as functions of different test temperatures and aluminum contents; (b) the macroscopic samples after impact [21].

http://dx.doi.org/10.5772/intechopen.82351

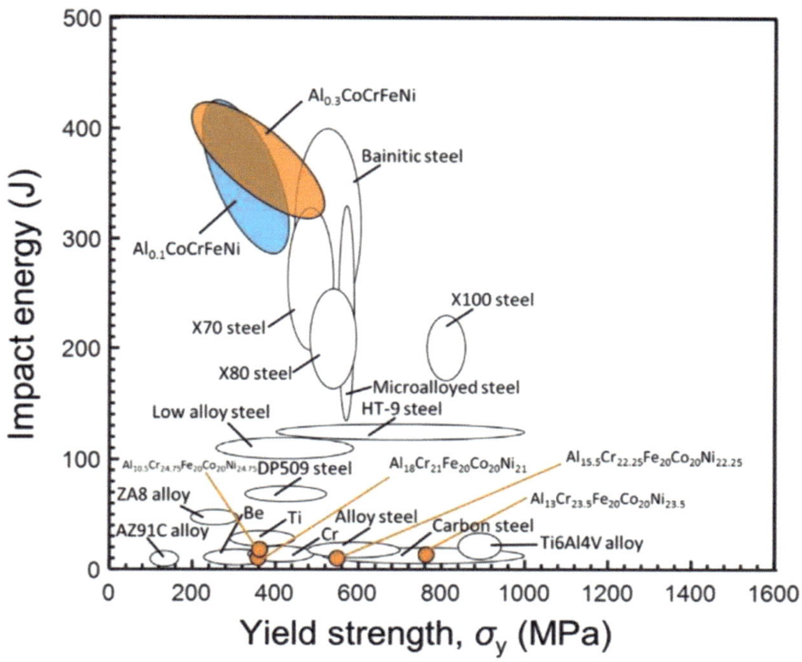

Figure 8. Map of impact energy versus yield strength for different materials [25].

In all these alloys, just as **Figures 6** and **7(a)** show, no obvious ductile-to-brittle transition (DBT) occurs as in many such conventional alloys as steels, amorphous alloys, Mg alloys, porous metals, and nanocrystalline metals [22–24]. A great DBT property and a low DBT temperature are detrimental to cold condition applications. Nowadays, researchers have tried to reduce the DBT temperature of materials through various methods, and the emergence of these kinds of entropic alloys provides them with a good choice.

Furthermore, it can be seen from the impact energy-yield strength diagram shown in **Figure 8** [25] that the two kinds of entropic alloys $Al_{0.1}$ and $Al_{0.3}$ have excellent comprehensive mechanical properties of impact energy and yield strength. Although their yield strength is relatively smaller than that of some carbon steels, their impact toughness is several times to several tens of times higher than that of various steels and titanium alloys.

3. Deformation mechanism at low temperature

3.1. Synergistic effect of dislocations and nano-twins

Stacking fault is a lattice defect that only changes the sub-neighbor relation of atoms and hardly produces lattice distortion. Usually, the energy added by the unit area fault is called the stacking-fault energy, which is an intrinsic property of metals. The smaller the stacking-fault energy, the greater the probability of stacking faults occur. In low-stacking-fault-energy alloys, such as austenitic steels and magnesium alloys, dissociation into partial dislocations is more energetically favorable, and the spacing between the partial dislocations is larger. As the spacing increases, cross slip and climb become more difficult, increasing strength. Low-stacking-fault-energy alloys are also more likely to deform by twinning, increasing dislocation storage capacity, strain-hardening rate, and ductility.

Huang et al. found that the entropic alloys have lower-stacking-fault energy and are particularly effective in low-temperature environments [26]. Based on this, it can be inferred that the entropic alloys will undergo twinning at low temperatures and have excellent performance, which has been confirmed by the subsequent studies.

A large number of experiments have studied the nature of the excellent low-temperature properties of entropic alloys from various aspects. It is now believed that the synergistic effect of various deformation mechanisms and the inhibition of crack growth by nano-twins are the essence of excellent mechanical properties [27]. In the room temperature, the deformation is mainly realized through dislocation, and the proliferation and expansion of large numbers of dislocations make the entropic alloys have excellent plasticity. The dimple diameter is almost entirely micron size (1–10 μm). While in the low temperature, nano-twins appear and interact with dislocations. Among different grains, the twin exhibit strong synergistic deformation, while a large number of cross twisting appear in the same grain. The interaction of the oriented twins greatly refines the grains, and the nanoscale dimples appear in the fracture. The combination of these factors makes the entropic alloy exhibit excellent low-temperature performance.

3.2. Phase transition

Further in-depth studies have found that there are small amounts of phase changes in the deformed sample, that is, the FCC to HCP phase transition occurs when the material is deformed in the liquid nitrogen and liquid helium temperature zones. In the ultralow temperature, the FCC to HCP transition fully demonstrates that HCP is a stable phase at low temperature and the FCC structure is a metastable phase, which is similar to the deformation behavior of 316LN in ultralow temperature. However, differences are that the volume fraction of 316LN in the phase transition is about 24%, while the entropic alloys only have a very small amount of FCC phase transformed. This also indirectly indicates that the entropic alloys have a more stable FCC structure than the conventional materials due to their high mixing entropy.

The results of TEM and HRTEM show that the dislocation decomposition produces partial dislocations, and the slip of them in every other layer of the close-packed planes causes the phase transition. Due to the special orientation relationship between FCC and HCP, coupled with the promotion of low temperature, atoms in every other 111 layers which stack in the order of ABCABC form glide with Shockley partial dislocation, which in turn cause the appearance of HCP structure as the stacking order of ABAB.

In addition, although the phase transformation is beneficial to improve the strength and plasticity, the stress concentrations are also prone to occur in the boundary, and the occurrence of twinning hinders the dislocation motion. Therefore, the plasticity is lower than that at 77 K.

4. Applications and prospects

With the continuous development of science and technology, low-temperature technology has played an increasingly important role in human production since its discovery, especially in aerospace science, biology, and life sciences. In the low-temperature state (including extremely low temperature), the properties of the substance may change dramatically: gas liquefaction or solidification, slow or stop of biological cell metabolism to prolong its life, loss of electrical resistance of the conductor to form a superconductor, and the like. Therefore, both the development of refrigeration cryogenic equipment and the application of ultralow-temperature materials have greatly promoted the development of science and technology and greatly improved the level of human production, especially in the treatment of major human diseases and the exploration of outer space, such as the development of LNG cryogenic carriers and low-temperature superconductors. The development of cryogenic engineering relies on the development and renewal of low-temperature materials.

A large number of experimental studies have shown that the entropic alloys have excellent low-temperature mechanical properties. It has been investigated that these great mechanical properties far exceed those of some high-alloying austenitic stainless steel and they have great potentials for low-temperature applications, including low-temperature storage and low-temperature structural parts. The researches on the deformation mechanisms are also becoming more and more mature. However, there are still some problems that need to be resolved. Firstly, the low-temperature performance of entropic alloys in different states, different deformation levels, and different heat treatment processes also requires systematic research. Secondly, when serving at low temperatures, not only excellent mechanical properties are required, but also corrosion resistance and easy processing ability are needed for the alloys, which still need to be further investigated. Furthermore, another interesting phenomenon is that the stress-strain curve appears to be serrated in a specific temperature and strain-rate range [28, 29], which will seriously affect the properties and applications, because the unsmooth stress-strain curve is difficult to predict the material safety factor. As a result, it is necessary to study the serration behaviors more thoroughly.

5. Conclusion

The entropic alloys have excellent low-temperature mechanical properties, including high tensile strength, high fracture toughness, and high impact resistance, some of which even exceeds the limits of existing materials. The main reason for its excellent performance is that the ultralow-temperature environment and high-entropy effect are beneficial to reduce the stacking-fault energy, which makes it easy of deformation mechanism transition from dislocation dominant to twining dominant. The appearance of cross twisting and synergistic twinning greatly improves the low-temperature properties. In addition, the local FCC to HCP phase transitions also affect their comprehensive properties.

Acknowledgements

Y. Zhang would like to thank the financial support from National Natural Science Foundation of China (NSFC), Grant No. 51471025 and 51671020.

Conflict of interest

The authors declare that they have no conflict of interest.

Author details

Rui Xuan Li and Yong Zhang*

*Address all correspondence to: drzhangy@ustb.edu.cn

State Key Laboratory for Advanced Metals and Materials, University of Science and Technology Beijing (USTB), Beijing, China

References

[1] Yeh JW, Chen SK, Lin SJ, Gan JY, Chin TS, Shun TT, et al. Nanostructured high-entropy alloys with multiple principal elements: Novel alloy design concepts and outcomes. Advanced Engineering Materials. 2004;**6**(5):299-303

[2] Zhang W, Liaw PK, Zhang Y. Science and technology in high-entropy alloys. Science China Materials. 2018;**61**(1):2-22

[3] Zhang Y, Zuo TT, Tang Z, Gao MC, Dahmen KA, Liaw PK, et al. Microstructures and properties of high-entropy alloys. Progress in Materials Science. 2014;**61**:1-93

[4] Yeh JW, Chen YL, Lin SJ, Chen SK. High-entropy alloys—A new era of exploitation. Materials Science Forum. 2007;**560**:1-9

[5] Wu Z, Bei H, Pharr GM, George EP. Temperature dependence of the mechanical properties of equiatomic solid solution alloys with face-centered cubic crystal structures. Acta Materialia. 2014;**81**:428-441

[6] Lilensten L, Couzinié JP, Bourgon J, Perrière L, Dirras G, Prima F, et al. Design and tensile properties of a bcc Ti-rich high-entropy alloy with transformation-induced plasticity. Materials Research Letters. 2016;**5**(2):110-116

[7] Tracy CL, Park S, Rittman DR, Zinkle SJ, Bei H, Lang M, et al. High pressure synthesis of a hexagonal close-packed phase of the high-entropy alloy CrMnFeCoNi. Nature Communications. 2017;**8**:15634

[8] Yeh JW, Chang SY, Hong YD, Chen SK, Lin SJ. Anomalous decrease in X-ray diffraction intensities of Cu–Ni–Al–Co–Cr–Fe–Si alloy systems with multi-principal elements. Materials Chemistry and Physics. 2007;**103**(1):41-46

[9] Yang X, Zhang Y. Prediction of high-entropy stabilized solid-solution in multi-component alloys. Materials Chemistry and Physics. 2012;**132**(2-3):233-238

[10] Li D, Li C, Feng T, Zhang Y, Sha G, Lewandowski JJ, et al. High-entropy $Al_{0.3}CoCrFeNi$ alloy fibers with high tensile strength and ductility at ambient and cryogenic temperatures. Acta Materialia. 2017;**123**:285-294

[11] Youssef KM, Zaddach AJ, Niu C, Irving DL, Koch CC. A novel low-density, high-hardness, high-entropy alloy with close-packed single-phase nanocrystalline structures. Materials Research Letters. 2014;**3**(2):95-99

[12] Xia S, Gao MC, Yang T, Liaw PK, Zhang Y. Phase stability and microstructures of high entropy alloys ion irradiated to high doses. Journal of Nuclear Materials. 2016;**480**:100-108

[13] Senkov ON, Wilks GB, Miracle DB, Chuang CP, Liaw PK. Refractory high-entropy alloys. Intermetallics. 2010;**18**(9):1758-1765

[14] Poletti MG, Fiore G, Gili F, Mangherini D, Battezzati L. Development of a new high entropy alloy for wear resistance: $FeCoCrNiW_{0.3}$ and $FeCoCrNiW_{0.3}$+5at.% of C. Materials and Design. 2017;**115**:247-254

[15] Hemphill MA, Yuan T, Wang GY, Yeh JW, Tsai CW, Chuang A, et al. Fatigue behavior of $Al_{0.5}CoCrCuFeNi$ high entropy alloys. Acta Materialia. 2012;**60**(16):5723-5734

[16] Gali A, George EP. Tensile properties of high- and medium-entropy alloys. Intermetallics. 2013;**39**:74-78

[17] Otto F, Dlouhý A, Somsen C, Bei H, Eggeler G, George EP. The influences of temperature and microstructure on the tensile properties of a CoCrFeMnNi high-entropy alloy. Acta Materialia. 2013;**61**(15):5743-5755

[18] Gludovatz B, Hohenwarter A, Catoor D, Chang EH, George EP, Ritchie RO. A fracture-resistant high-entropy alloy for cryogenic applications. Science. 2014;**345**(6201):1153-1158

[19] Gludovatz B, Hohenwarter A, Thurston KV, Bei H, Wu Z, George EP, et al. Exceptional damage-tolerance of a medium-entropy alloy CrCoNi at cryogenic temperatures. Nature Communications. 2016;**7**:10602

[20] Li D, Zhang Y. The ultrahigh charpy impact toughness of forged $Al_xCoCrFeNi$ high entropy alloys at room and cryogenic temperatures. Intermetallics. 2016;**70**:24-28

[21] Xia SQ, Gao MC, Zhang Y. Abnormal temperature dependence of impact toughness in Al_xCoCrFeNi system high entropy alloys. Materials Chemistry and Physics. 2018;**210**:213-221

[22] Sokolov MA, Tanigawa H, Odette GR, Shiba K, Klueh RL. Fracture toughness and Charpy impact properties of several RAFMS before and after irradiation in HFIR. Journal of Nuclear Materials. 2007;**367-370**:68-73

[23] Li H, Ebrahimi F. Ductile-to-brittle transition in nanocrystalline metals. Advanced Materials. 2005;**17**(16):1969-1972

[24] Raghavan R, Murali P, Ramamurty U. On factors influencing the ductile-to-brittle transition in a bulk metallic glass. Acta Materialia. 2009;**57**(11):3332-3340

[25] Li W, Liaw PK, Gao Y. Fracture resistance of high entropy alloys: A review. Intermetallics. 2018;**99**:69-83

[26] Huang S, Li W, Lu S, Tian F, Shen J, Holmström E, et al. Temperature dependent stacking fault energy of FeCrCoNiMn high entropy alloy. Scripta Materialia. 2015;**108**:44-47

[27] Zhang Z, Mao M, Wang J, Gludovatz B, Zhang Z, Mao SX, et al. Nanoscale origins of the damage tolerance of the high-entropy alloy CrMnFeCoNi. Nature Communications. 2015;**6**:10143

[28] Chen S, Xie X, Li W, Feng R, Chen B, Qiao J, et al. Temperature effects on the serrated behavior of an $Al_{0.5}$CoCrCuFeNi high-entropy alloy. Materials Chemistry and Physics. 2018;**210**:20-28

[29] Zhang Y, Liu JP, Chen SY, Xie X, Liaw PK, Dahmen KA, et al. Serration and noise behaviors in materials. Progress in Materials Science. 2017;**90**:358-460